Power Systems

More information about this series at http://www.springer.com/series/4622

Md. Abdus Salam · Quazi M. Rahman

Power Systems Grounding

 Springer

Md. Abdus Salam
Universiti Teknologi Brunei
Bandar Seri Begawan
Brunei Darussalam

Quazi M. Rahman
University of Western Ontario
London, ON
Canada

ISSN 1612-1287
Power Systems
ISBN 978-981-10-9165-0
DOI 10.1007/978-981-10-0446-9

ISSN 1860-4676   (electronic)

ISBN 978-981-10-0446-9   (eBook)

Printed on acid-free paper

This Springer imprint is published by Springer Nature
The registered company is Springer Science+Business Media Singapore Pte Ltd.

To all my teachers and well-wishers who have helped me grow professionally over the years

Md. Abdus Salam

In loving memory of my beloved mother Mrs. Matiza Begum, a wonderful teacher and a mentor, whose utmost selfless love and prayers for me to Almighty Allah have put me in a position where, I'm today. I appreciate the opportunity I've had as a result of the hard work and sacrifice of both of my parents

Quazi M. Rahman

Preface

The book *Power System Grounding* is intended for both lower- and upper-level undergraduate students studying power system, design, and measurement of grounding system, as well as a reference for power system engineers. For reference, this book has been written in a step-by-step method. In this method, this book covers the fundamental knowledge of power, transformer, different types of faults, soil properties, soil resistivity, and ground resistance measurement methods. This book also includes fundamental and advanced theories related to the grounding system.

Nowadays, the demand for smooth, safe, and reliable power supply is increasing due to an increase in the development of residential, commercial, and industrial sectors. The safe and reliable power supply system is interrupted due to different faults including lightning, short circuit, and ground faults. A good grounding system can protect substations, and transmission and distribution networks from these kinds of faults. In addition, a good grounding system ensures the safety of humans in the areas of faulty substations in case of ground faults, and decreases the electromagnetic interference in substations.

This book is organized into seven chapters and two appendices. Chapter 1 deals with the fundamental knowledge of power analysis. Transformer fundamentals and practices are discussed in Chap. 2. Chapter 3 covers the issues on the symmetrical and unsymmetrical faults. Chapter 4 includes the grounding system parameters and resistance. In this chapter, different parameters related to the grounding system and expressions of resistance with different sizes of electrodes are discussed. Chapter 5 presents ideas on different types of soil, properties of soil, influence of different parameters on soil, current density, and Laplace and Poisson equations and their solutions. Chapter 6 describes different measurement methods of soil resistivity. The grounding resistance measurement methods are discussed in Chap. 7.

This book will offer both students and practice engineers the fundamental concepts in conducting practical measurements on soil resistivity and ground resistance at residential and commercial areas, substations, and transmission and distribution networks.

Authors would like to thank Fluke Corporation for providing high-quality photographs and equipments Fluke 1625, Fluke 1630, and some technical contents.

Special thanks go to our Ex-MSc Engineering student Luis Beltran, The University of Western Ontario, London, Canada who developed part of the derivation on the measurement of ground resistance.

The authors would like to express their sincere thanks to the production staff of Springer Publishing Company for their continuous help with the preparation of the manuscript and bringing this book to completion.

Lastly, we express our heartfelt thanks to our respective spouses and children for their continued patience during the preparation of the book.

Contents

Chapter 1
Power Analysis

1.1 Introduction

Electrical power is the time rate of receiving or delivering electrical energy that depends on the voltage and current quantities. In an ac circuit, the current and voltage quantities vary with time, and so the electrical power. Every electrical device such as ceiling fan, bulb, television, iron, micro-wave oven, DVD player, water-heater, refrigerator, etc. has a power rating that specifies the amount of power that device requires to operate. An electrical equipment with high power rating generally draws large amount of current from the energy source (e.g., voltage source) which increases the energy consumption. Nowadays, Scientists and Engineers are jointly working on the design issues of the electrical equipment to reduce the energy consumption. Since in the design stage, power analysis plays a vital role, a clear understanding on power analysis fundamentals becomes a most important prep work for any electrical engineer. This chapter reviews these fundamental concepts that include instantaneous power, average power, complex power, power factor, power factor correction and three-phase power.

1.2 Instantaneous Power

The instantaneous power (in watt) $p(t)$ is defined as the product of time varying voltage $v(t)$ and current $i(t)$, and it is written as [1],

$$p(t) = v(t) \times i(t) \qquad (1.1)$$

© Springer Science+Business Media Singapore 2016
Md.A. Salam and Q.M. Rahman, *Power Systems Grounding*,
Power Systems, DOI 10.1007/978-981-10-0446-9_1

Consider that the time varying excitation voltage $v(t)$ for an ac circuit is given by,

$$v(t) = V_m \sin(\omega t + \phi) \tag{1.2}$$

where, ω is the angular frequency and ϕ is the phase angle associated to the voltage source. In this case, the expression of the resulting current $i(t)$ in an ac circuit as shown in Fig. 1.1 can be written as,

$$i(t) = \frac{v(t)}{Z|\phi} \tag{1.3}$$

where, $Z|\phi$ is the impedance of the circuit in polar form, in which Z is the magnitude of the circuit impedance.

Substituting Eq. (1.2) into Eq. (1.3) yields,

$$i(t) = \frac{V_m \sin(\omega t + \phi)}{Z|\phi} = \frac{V_m|\phi}{Z|\phi} = \frac{V_m}{Z}|0° \tag{1.4}$$

$$i(t) = I_m \sin \omega t \tag{1.5}$$

where, the maximum current is,

$$I_m = \frac{V_m}{Z} \tag{1.6}$$

Substituting Eqs. (1.2) and (1.5) into Eq. (1.1) yields,

$$p(t) = V_m \sin(\omega t + \phi) \times I_m \sin \omega t \tag{1.7}$$

$$p(t) = \frac{V_m I_m}{2} \times 2 \sin(\omega t + \phi) \sin \omega t \tag{1.8}$$

Fig. 1.1 A simple ac circuit

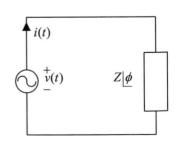

$$p(t) = \frac{V_m I_m}{2} [\cos \phi - \cos(2\omega t + \phi)] \tag{1.9}$$

$$p(t) = \frac{V_m I_m}{2} \cos \phi - \frac{V_m I_m}{2} \cos(2\omega t + \phi) \tag{1.10}$$

Equation (1.10) presents the resultant instantaneous power.

Example 1.1 An ac series circuit is having the excitation voltage and the impedance $v(t) = 5\sin(\omega t - 25°)$ V and $Z = 2\lfloor 15°$ Ω, respectively. Determine the instantaneous power.

Solution

The value of the series current is,

$$I = \frac{5\lfloor -25°}{2\lfloor 15°} = 2.5\lfloor -40° \text{ A}$$
$$i(t) = 2.5\sin(\omega t - 40°) \text{ A}$$

The instantaneous power can be determined as,

$$p(t) = \frac{5 \times 2.5}{2} \times 2\sin(\omega t - 25°)\sin(\omega t - 40°)$$
$$p(t) = \frac{5 \times 2.5}{2} [\cos 15° - \cos(2\omega t - 65°)]$$
$$p(t) = 6.04 - 6.25\cos(2\omega t - 65°) \text{ W}$$

Practice problem 1.1
The impedance and the current in an ac circuit are given by $Z = 2.5\lfloor 30°$ Ω and $i(t) = 3\sin(314t - 15°)$ A, respectively. Calculate the instantaneous power.

1.3 Average and Apparent Power

Energy consumption of any electrical equipment depends on the power rating of the equipment and the duration of its operation, and this brings in the idea of average power. The average power (in watt) is defined as the average of the instantaneous power over one periodic cycle, and is given by [2, 3],

$$P = \frac{1}{T} \int_0^T p(t)dt \tag{1.11}$$

Substituting Eq. (1.10) into Eq. (1.11) yields,

$$P = \frac{V_m I_m}{2T} \int_0^T [\cos \phi - \cos(2\omega t + \phi)] \, dt \tag{1.12}$$

$$P = \frac{V_m I_m}{2T} \cos \phi \, [T] - \frac{V_m I_m}{2T} \int_0^T \cos(2\omega t + \phi) \, dt \tag{1.13}$$

$$P = \frac{V_m I_m}{2} \cos \phi - \frac{V_m I_m}{2T} \int_0^T \cos(2\omega t + \phi) \, dt \tag{1.14}$$

The second term of Eq. (1.14) is a cosine wave, and the average value of the cosine wave over one cycle is zero. Therefore, the final expression of the average power becomes,

$$P = \frac{V_m I_m}{2} \cos \phi \tag{1.15}$$

Sometimes, average power is referred to as the true or real power expressed in watts which is the actual power dissipated by the load.

Equation (1.15) can be rearranged as,

$$P = \frac{V_m}{\sqrt{2}} \frac{I_m}{\sqrt{2}} \cos \phi \tag{1.16}$$

$$P = V_{rms} I_{rms} \cos \phi \tag{1.17}$$

where, V_{rms} and I_{rms} are the root-mean-square (rms) values of the voltage and current components, respectively. The product $V_{rms} I_{rms}$ is known as the apparent power expressed as volt-ampere (VA).

The phase angle between voltage and current quantities associated to a resistive circuit is zero since they are in phase with each other. From Eq. (1.17), the average power associated to a resistive circuit component can be written as,

$$P_R = \frac{V_m I_m}{2} \cos 0° \tag{1.19}$$

$$P_R = \frac{V_m I_m}{2} \tag{1.20}$$

Applying Ohm's law to Eq. (1.20) yields,

$$P_R = \frac{V_m I_m}{2} = \frac{1}{2} I_m^2 R = \frac{V_m^2}{2R} \qquad (1.21)$$

The voltage and current quantities associated to either an inductive or a capacitive circuit element are always 90 degree out of phase with each other, and this results in a zero average power (P_L or P_C) as shown in the following expression,

$$P_L = P_C = \frac{V_m I_m}{2} \cos 90^\circ = 0 \qquad (1.22)$$

Example 1.2 Figure 1.2 shows an electric circuit with different elements. Determine the total average power supplied by the source and absorbed by each elements of the circuit.

Solution

The equivalent impedance of the circuit is,

$$Z_t = 2 + \frac{3 \times (4 + j6)}{7 + j6} = 4.31\underline{|8.48^\circ}\ \Omega$$

The value of the source current is,

$$I = \frac{15\underline{|45^\circ}}{4.31\underline{|8.48^\circ}} = 3.48\underline{|36.52^\circ}\ A$$

The current through the 4 Ω resistor is,

$$I_1 = \frac{3.48\underline{|36.52^\circ} \times 3}{7 + j6} = 1.13\underline{|-4.08^\circ}\ A$$

The current through the 3 Ω resistor is,

$$I_2 = \frac{3.48\underline{|36.52^\circ} \times (4 + j6)}{7 + j6} = 2.72\underline{|52.23^\circ}\ A$$

Fig. 1.2 An electrical circuit

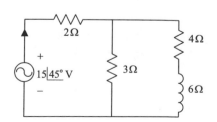

Fig. 1.3 A series ac circuit

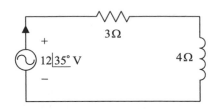

The average power absorbed by the $3\,\Omega$ resistor is,

$$P_1 = \frac{1}{2} \times 2.72^2 \times 3 = 11.1\,\text{W}$$

The average power absorbed by the $4\,\Omega$ resistor is,

$$P_2 = \frac{1}{2} \times 1.13^2 \times 4 = 2.55\,\text{W}$$

The average power absorbed by the $2\,\Omega$ resistor is,

$$P_3 = \frac{1}{2} \times 3.48^2 \times 2 = 12.11\,\text{W}$$

The average power supplied by the source is,

$$P_4 = \frac{1}{2} \times 15 \times 3.48 \times \cos(45° - 36.52°) = 25.81\,\text{W}$$

The total average power absorbed by the elements is,

$$P_5 = 11.1 + 2.55 + 12.11 = 25.76\,\text{W}$$

Practice problem 1.2
Find the average power for each element of the circuit shown in Fig. 1.3.

1.4   Power Factor

Assuming that the phase angles associated to the voltage and current components
are θ_v and θ_i respectively, the average power given in Eq. (1.17) can be written as,

$$P = V_{rms}I_{rms}\cos(\theta_v - \theta_i) \tag{1.23}$$

From Eq. (1.23) the power factor (pf) can be introduced as,

$$pf = \frac{P}{V_{rms}I_{rms}} = \cos(\theta_v - \theta_i) = \cos\phi \tag{1.24}$$

So, the power factor (pf) is defined as the ratio of the average power to the apparent power associated to a circuit element. Also, based on the mathematical expression in equation (1.50) the power factor can be defined as the cosine of the angle ϕ which is the resultant phase difference between the voltage-phase (θ_v) and the current phase (θ_i) associated to a circuit component. The angle ϕ is often referred to as power factor angle.

The power factor in a circuit element is considered as lagging when the current lags the voltage, whereas, it is considered as leading when the current leads the voltage. In general, industrial loads are inductive and so they have a lagging power factors. A capacitive load has a leading power factor. Every industry always maintains a required power factor by using a power factor improvement unit. It is economically viable for an industry to have a unity power factor or a power factor as close to unity. Few disadvantages of having a load wih low power factor are (i) the kVA rating of the electrical machines is increased, (ii) larger conductor size is required to transmit or distribute electric power at constant voltage, (iii) copper losses are increased, and (iv) voltage regulation is smaller.

1.5 Complex Power and Reactive Power

The product of the rms voltage-phasor and the conjugate of the rms current-phasor associated to an electrical component is known as the complex power (in volt-ampere or VA), and it is denoted by S. Mathematically, it can be written as,

$$\mathbf{S} = V_{rms}I_{rms}^* \tag{1.25}$$

Considering that the phase angles associated to the voltage and current components are θ_v and θ_i respectively, Eq. (1.25) can be written as,

$$\mathbf{S} = V_{rms}\angle\theta_v I_{rms}\angle -\theta_i = V_{rms}I_{rms}\angle\theta_v - \theta_i \tag{1.26}$$

and this becomes,

$$\mathbf{S} = V_{rms}I_{rms}\cos(\theta_v - \theta_i) + jV_{rms}I_{rms}\sin(\theta_v - \theta_i) = P + jQ \tag{1.27}$$

where,

$$P = \mathrm{Re}(\mathbf{S}) = V_{rms}I_{rms}\cos(\theta_v - \theta_i) \tag{1.28}$$

$$Q = \mathrm{Im}(\mathbf{S}) = V_{rms}I_{rms}\sin(\theta_v - \theta_i) \tag{1.29}$$

Fig. 1.4 A circuit with
impedance

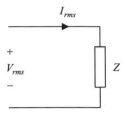

As shown in Eq. (1.27), the real part of the complex power is the real or average power (introduced in the previous section) while the imaginary part of the complex power is known as the reactive power. The reactive power is expressed as volt-ampere-reactive (VAR).

Now, let's discuss the concept of complex power and reactive power with the aid of the above equations, and circuit diagram shown in Fig. 1.4. The impedance of this circuit is given by,

$$\mathbf{Z} = R + jX \tag{1.30}$$

where, R is the resistive component and X is the reactive component of the circuit.

The rms value of the current is,

$$\mathbf{I}_{rms} = \frac{\mathbf{V}_{rms}}{\mathbf{Z}} \tag{1.31}$$

Substituting Eq. (1.31) into Eq. (1.25) yields,

$$\mathbf{S} = \mathbf{V}_{rms}\frac{\mathbf{V}_{rms}^*}{\mathbf{Z}^*} = \frac{V_{rms}^2}{\mathbf{Z}^*} \tag{1.32}$$

Again, with the aid of Eq. (1.25), Eq. (1.31) can be rearranged as,

$$\mathbf{S} = \mathbf{I}_{rms}\mathbf{Z}\mathbf{I}_{rms}^* = I_{rms}^2\mathbf{Z}^* \tag{1.33}$$

Substituting Eq. (1.30) into Eq. (1.28) yields,

$$\mathbf{S} = I_{rms}^2(R + jX) = I_{rms}^2 R + jI_{rms}^2 X = P + jQ \tag{1.34}$$

where, P is the real power and Q is the reactive power, and these quantities are expressed as,

$$P = \mathrm{Re}(\mathbf{S}) = I_{rms}^2 R \tag{1.35}$$

$$Q = \mathrm{Im}(\mathbf{S}) = I_{rms}^2 X \tag{1.36}$$

In this case, by comparing Eqs. (1.29) and (1.36), we can conclude that the reactive power is the energy that is traded between the source and the reactive part of the load. It is worth noting that the magnitude of the complex power is the apparent power which has been introduced in the previous section.

Now, let's look at the variation of the complex power in terms of the characteristics of the circuit components. In case of resistive circuit component, the expression of complex power becomes,

$$\mathbf{S}_R = P_R + jQ_R = I_{rms}^2 R \tag{1.37}$$

Here, the real power and the reactive power quantities are,

$$P_R = I_{rms}^2 R \tag{1.38}$$

$$Q = 0 \tag{1.39}$$

The complex power for an inductive component is,

$$\mathbf{S}_L = P_L + jQ_L = jI_{rms}^2 X_L \tag{1.40}$$

Here, the real power and the reactive power quantities are,

$$P_L = 0 \tag{1.41}$$

$$Q_L = I_{rms}^2 X_L \tag{1.42}$$

Similarly, the complex power for a capacitive component is,

$$\mathbf{S}_C = P_C + jQ_C = -jI_{rms}^2 X_C \tag{1.43}$$

Here, the real power and the reactive power quantities are,

$$P_C = 0 \tag{1.44}$$

$$Q_C = I_{rms}^2 X_C \tag{1.45}$$

Now we can introduce another useful power-analysis quantity by dividing Eq. (1.29) by Eq. (1.28) yields,

$$\frac{Q}{P} = \tan(\theta_v - \theta_i) = \tan\theta \tag{1.46}$$

The relationship between the power factor angle to P and Q is known as power triangle. Similarly, we can find the relationship between different components in an impedance which is called impedance triangle. Figure 1.5 shows the power and

Fig. 1.5 Power and impedance triangles for lagging power factor

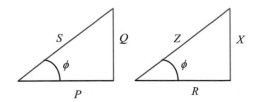

impedance triangles for lagging power factor. The following points are noted for different power factors.

$Q = 0$ for resistive loads i.e., unity power factor,
$Q > 0$ for inductive loads i.e., lagging power factor,
$Q < 0$ for capacitive loads i.e., leading power factor.

Example 1.3 Determine the source current, apparent power, real power and reactive power of the circuit shown in Fig. 1.6.

Solution

The rms value of the voltage is,

$$V_{rms} = 5 \times \sqrt{2} \,\underline{|35°} = 7.07\underline{|35°}\ \text{V}$$

The value of the inductive reactance is,

$$X_L = 2 \times 4 = 8\,\Omega$$

The value of the circuit impedance is,

$$Z = 5 + j8 = 9.43\underline{|58°}\ \Omega$$

The value of the source current is,

$$I_{rms} = \frac{7.07\underline{|35°}}{9.43\underline{|58°}} = 0.75\underline{|-23°}\ \text{A}$$

Fig. 1.6 Circuit with resistance and inductance

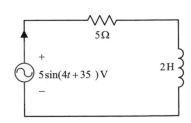

The value of the complex power is,

$$S = V_{rms}I^*_{rms} = 7.07\underline{|35°} \times 0.75\underline{|23°} = 2.81 + j4.5 \text{ W}$$

The value of the apparent power is,

$$P_A = |S| = 5.305 \text{ VA}$$

The value of the real power is,

$$P = \text{Re}(S) = 2.81 \text{ W}$$

The value of the reactive power is,

$$Q = \text{Im}(S) = 4.5 \text{ VAR}$$

Practice problem 1.3
The excitation voltage, resistor and inductance of a series circuit are $v(t) = 6\sqrt{2}\sin(15t + 45°)$ V, $6\,\Omega$ and 1.5 H, respectively. Find the source current, apparent power, real power and reactive power.

1.6   Complex Power Balance

Two loads connected in parallel with a voltage source is shown in Fig. 1.7. According to the conservation of energy, the total real power supplied by the source is equal to the sum of the real powers absorbed by the load. Similarly, the total complex power supplied by the source is equal to the sum of the complex powers delivered to each load. Here, the source current can be expressed as,

$$\mathbf{I} = \mathbf{I}_1 + \mathbf{I}_2 \tag{1.47}$$

Considering that V and I are the rms quantities, the complex power can be written as,

$$\mathbf{S} = \mathbf{VI}^* \tag{1.48}$$

Fig. 1.7 Parallel loads

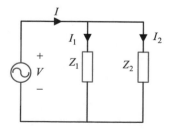

Substituting Eq. (1.41) into Eq. (1.42) yields,

$$\mathbf{S} = \mathbf{V}[\mathbf{I}_1 + \mathbf{I}_2]^* = \mathbf{VI}_1^* + \mathbf{VI}_2^* \tag{1.49}$$

Example 1.4 An ac circuit is shown in Fig. 1.8. Find the source current, power absorbed by each load and the total complex power. Assume that the source voltage is given in rms value.

Solution

The current in the first and second branches are determined as,

$$I_1 = \frac{25}{4} = 6.25 \text{ A}$$

$$I_2 = \frac{25}{3 + j4} = 5\underline{|-53.13°} \text{ A}$$

Then, the value of the source current is calculated as,

$$I = 6.25 + 5\underline{|-53.13°} = 10.08\underline{|-23.39°} \text{ A}$$

The value of the complex power for the first branch is,

$$S_1 = VI_1^* = 25 \times 6.25 = 156.25 \text{ W} + j0 \text{ VAR}$$

The value of the complex power for the second branch is,

$$S_2 = VI_2^* = 25 \times 5\underline{|53.13°} = 75 \text{ W} + j100 \text{ VAR}$$

The value of the total complex power is,

$$S = S_1 + S_2 = 231 \text{ W} + j100 \text{ VAR}$$

Which can also be evaluated as,

$$S = VI^* = 25 \times 10.07\underline{|23.38°} = 231 \text{ W} + j100 \text{ VAR}$$

Fig. 1.8 A series-parallel ac circuit

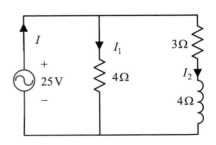

Fig. 1.9 A parallel ac circuit

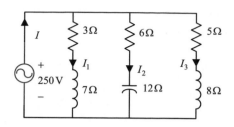

Practice problem 1.4

Determine the source current, the power absorbed by each load and the total complex power of an electrical circuit shown in Fig. 1.9.

1.7 Power Factor Correction

In the industry, inductive loads draw a lagging current which in turn increases the amount of reactive power. In this case, the kVA rating of the transformer and the size of the conductor should be increased to carry out the additional reactive power. Generally, capacitors are connected in parallel with the load to improve the low power factor by increasing the power factor value. Capacitor draws leading current, and partially or completely neutralizes the lagging reactive power of the load. Consider a single-phase inductive load as shown in Fig. 1.10. This load draws a lagging current I_1 at a power factor of $\cos \phi_1$ from the source.

A capacitor is connected in parallel with the load to improve the power factor as shown in Fig. 1.11. The capacitor will draw current that leads the source voltage by $90°$. The line current is the vector sum of the currents I_1 and I_2 as shown in Fig. 1.12. Figure 1.13 shows a power triangle to find the exact value of the capacitor. As shown in equation (na5), the reactive power of the original inductive load can be written as,

$$Q_1 = P \tan \phi_1 \qquad (1.50)$$

The main objective of this analysis is to improve the power factor from $\cos \phi_1$ to $\cos \phi_2$ ($\phi_2 < \phi_1$) without changing the real power. In this case, the expression of new reactive power will be,

Fig. 1.10 A single-phase inductive load

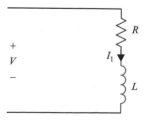

Fig. 1.11 Capacitor with a
parallel inductive load

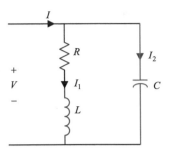

Fig. 1.12 Vector diagram for
power factor correction

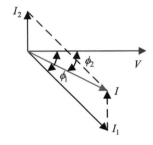

Fig. 1.13 Power triangle
with power factor correction

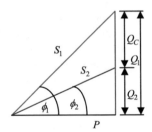

$$Q_2 = P \tan \phi_2 \tag{1.51}$$

The reduction in reactive power due to parallel capacitor is,

$$Q_C = Q_1 - Q_2 \tag{1.52}$$

Substituting Eqs. (1.50) and (1.51) into Eq. (1.52) yields,

$$Q_C = P(\tan \phi_1 - \tan \phi_2) \tag{1.53}$$

The reactive power due to capacitance is,

$$Q_C = \frac{V_{rms}^2}{X_C} = \omega C V_{rms}^2 \tag{1.54}$$

Substituting Eq. (1.54) into Eq. (1.53) yields,

$$Q_C = \omega C V_{rms}^2 = P(\tan \phi_1 - \tan \phi_2) \tag{1.55}$$

$$C = \frac{P(\tan \phi_1 - \tan \phi_2)}{\omega V_{rms}^2} \tag{1.56}$$

Equation (1.56) provides the value of the parallel capacitor.

Example 1.5 A 110 V(rms), 50 Hz power line is connected with 5 kW, 0.85 power factor lagging load. A capacitor is connected across the load to raise the power factor to 0.95. Find the value of the capacitance.

Solution

The value of the initial power factor is,

$$\cos \phi_1 = 0.85$$
$$\phi_1 = 31.79°$$

The value of the final power factor is,

$$\cos \phi_2 = 0.95$$
$$\phi_2 = 18.19°$$

The value of the parallel capacitor can be determined as,

$$C = \frac{5 \times 1000(\tan 31.79° - \tan 18.19°)}{2\pi \times 50 \times 110^2} = 383 \, \mu F$$

Practice problem 1.5
A 6 kW load with a lagging power factor of 0.8 is connected to a 125 V(rms), 60 Hz power line. A capacitor is connected across the load to improve the power factor to 0.95 lagging. Calculate the value of the capacitance.

Example 1.6 A load of 25 kW, 0.85 power factor lagging is connected across the line as shown in Fig. 1.14. Calculate the value of the capacitance when it is connected across the load to raise the power factor to 0.95. Also, Find the power

Fig. 1.14 Load with a line

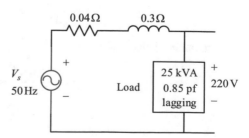

losses in the line before and after the capacitor is connected. Consider that the terminal voltage across the load is constant.

Solution

The initial power factor is,

$$\cos \phi_1 = 0.85$$
$$\phi_1 = 31.78°$$

The final power factor is,

$$\cos \phi_2 = 0.95$$
$$\phi_2 = 18.19°$$

The value of the power of the load is,

$$P = 25 \times 0.85 = 21.25 \, \text{kW}$$

The value of the capacitor is,

$$C = \frac{21.25 \times 1000(\tan 31.78° - \tan 18.19°)}{2\pi \times 50 \times 220^2} = 406.62 \, \mu\text{F}$$

Before adding capacitor:

The value of the line current is,

$$I_1 = \frac{21250}{0.85 \times 220} = 113.64 \, \text{A}$$

The power loss in the line is,

$$P_1 = 113.64^2 \times 0.04 = 516.53 \, \text{W}$$

After adding capacitor:

The value of the apparent power is,

$$S = \frac{21250}{0.95} = 22368.42 \, \text{VA}$$

The value of the line current is,

$$I_2 = \frac{22368.42}{220} = 101.67 \, \text{A}$$

Fig. 1.15 Load with a transmission line

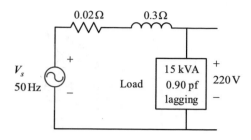

The power loss in the line is,

$$P_2 = 101.67^2 \times 0.04 = 413.51\,\text{W}$$

Practice problem 1.6

A 15 kVA load with 0.95 lagging power factor is connected across the line as shown in Fig. 1.15. Find the value of the capacitance when it is connected across the load to raise the power factor to 0.95. Also, determine the power losses in the line before and after the capacitor is connected. Assume that the terminal voltage across the load is constant.

1.8   Three-Phase System

AC generator generates three-phase sinusoidal voltages with constant magnitude but are displaced in phase by 120°. These voltages are called balanced voltages. In a three-phase generator, three identical coils a, b and c are displaced by 120° from each other. If the generator is turned by a prime mover then voltages V_{an}, V_{bn} and V_{cn} are generated. The balanced three-phase voltages and their waveforms are shown in Figs. 1.16 and 1.17, respectively. The expression of generated voltages can be represented as,

$$V_{an} = V_P \sin \omega t = V_P \underline{|0^\circ} \tag{1.57}$$

$$V_{bn} = V_P \sin(\omega t - 120^\circ) = V_P \underline{|-120^\circ} \tag{1.58}$$

Fig. 1.16 A balanced three-phase system

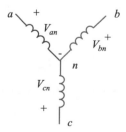

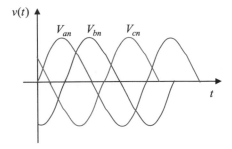

Fig. 1.17 Three-phase voltage waveforms

$$V_{cn} = V_P \sin(\omega t - 240°) = V_P \underline{|-240°} \qquad (1.59)$$

Here, V_P is the maximum voltage.

The magnitudes of phase voltages are the same, and these components can be expressed as,

$$|V_{an}| = |V_{bn}| = |V_{cn}| \qquad (1.60)$$

1.9 Naming Phases and Phase Sequence

The three-phase systems are denoted either by 1, 2, 3 or a, b, c. Sometimes, three phases are represented by three natural colours namely Red, Yellow and Blue, i.e., $R\ Y\ B$. If the generated voltages reach to their maximum or peak values in the sequential order of abc, then the generator is said to have a positive phase sequence as shown in Fig. 1.18a. If the generated voltages phase order is acb, then the

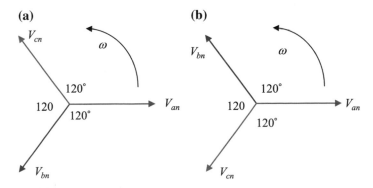

Fig. 1.18 a Positive sequence, **b** Negative sequence

generator is said to have a negative phase sequence which is shown in Fig. 1.18b. Here, the voltage, V_{an} is considered as the reference and the direction of rotation is considered as counterclockwise.

In the positive phase sequence, the passing sequence of voltages is given by $V_{an} - V_{bn} - V_{cn}$. In the negative phase sequence, the passing sequence of voltages is $V_{an} - V_{cn} - V_{bn}$.

1.10 Star Connection

The star-connection is often known as wye-connection. A generator is said to be a wye-connected generator when three connected coils form a connection as shown in Fig. 1.19. In this connection, one terminal of each coil is connected to a common point or neutral point n and the other three terminals represent the three-phase supply. The voltage between any line and the neutral point is called the phase voltage and are represented by V_{an}, V_{bn} and V_{cn} for phases a, b and c, respectively. The voltage between any two lines is called the line voltage. Line voltages between the lines a and b, b and c, c and a represented by V_{ab}, V_{bc} and V_{ca}, respectively.

Usually, the line voltage and the phase voltage are represented by V_L and V_P, respectively. In the Y-connection, the points to remember are (i) line voltage is equal to $\sqrt{3}$ times the phase voltage, (ii) line current is equal to the phase current and (iii) current (I_n) in the neutral wire is equal to the phasor sum of the three line currents. The neutral current is zero i.e., $I_n = 0$ for a balanced load.

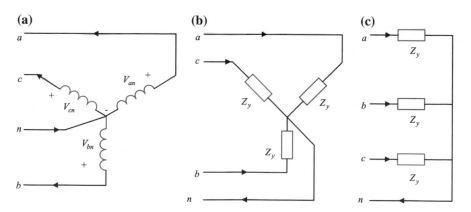

Fig. 1.19 Different types of three-phase wye connection. **a** Generator. **b** Load. **c** Load

1.11   Voltage and Current Relations for Y-Connection

Figure 1.20 shows a three-phase, Y-connected generator. Here V_{an}, V_{bn} and V_{cn} are the phase voltages and V_{ab}, V_{bc} and V_{ca} are the line voltages, respectively. The phase voltages for *abc* phase sequence are,

$$V_{an} = V_p \lfloor 0° \tag{1.61}$$

$$V_{bn} = V_p \lfloor -120° \tag{1.62}$$

$$V_{cn} = V_p \lfloor 120° \tag{1.63}$$

Apply KVL between lines *a* and *b* to the circuit in Fig. 1.21 and the equation is,

$$V_{an} - V_{bn} - V_{ab} = 0 \tag{1.64}$$

$$V_{ab} = V_{an} - V_{bn} \tag{1.65}$$

Substituting Eqs. (1.61) and (1.62) into Eq. (1.65) yields,

$$V_{ab} = V_P \lfloor 0° - V_P \lfloor -120° \tag{1.66}$$

$$V_{ab} = \sqrt{3} \, V_P \lfloor 30° \tag{1.67}$$

Similarly, the other line voltages can be derived from the appropriate loops of the circuit as shown in Fig. 1.20. The expressions of other line voltages are,

Fig. 1.20 Wye-connected generator

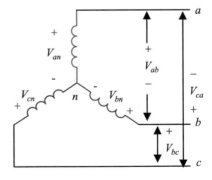

Fig. 1.21 Two lines of wye-connection

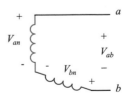

Fig. 1.22 Phasor diagram

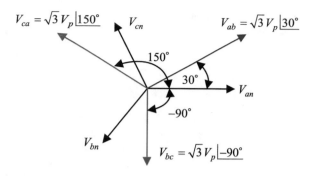

$$V_{bc} = V_{bn} - V_{cn} \tag{1.68}$$

$$V_{bc} = \sqrt{3}\ V_P\lfloor -90° \tag{1.69}$$

$$V_{ca} = V_{cn} - V_{an} \tag{1.70}$$

$$V_{bc} = \sqrt{3}\ V_P\lfloor 150° \tag{1.71}$$

Figure 1.22 shows the relationship between the phase voltages and the line voltages. From Eqs. (1.67), (1.69) and (1.71), it can be concluded that the magnitude of the line voltage is equal to $\sqrt{3}$ times the magnitude of the phase voltage. The general relationship between the line voltage and the phase voltage is,

$$V_L = \sqrt{3}\ V_P \tag{1.72}$$

From Fig. 1.20, it is seen that the phase current is equal to the line current and this relationship is,

$$I_L = I_P \tag{1.73}$$

Example 1.7 Figure 1.23 shows a three-phase, Y-connected generator. The generator generates equal magnitude of phase voltage of 200 V rms. For *abc* phase sequence, write down the phase voltages and the line voltages.

Solution

The phase voltages are,

$$V_{an} = 200\lfloor 0°\ \text{V}$$
$$V_{bn} = 200\lfloor -120°\ \text{V}$$
$$V_{cn} = 200\lfloor 120°\ \text{V}$$

Fig. 1.23 A Y-connection generator

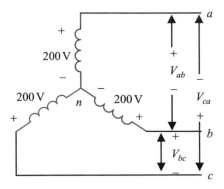

The line voltages are,

$$V_{ab} = \sqrt{3} \times 200\underline{|30°} = 346.40\underline{|30°}\ \text{V}$$
$$V_{bc} = \sqrt{3} \times 200\underline{|-120° + 30°} = 346.40\underline{|-90°}\ \text{V}$$
$$V_{ca} = \sqrt{3} \times 200\underline{|120° + 30°} = 346.40\underline{|150°}\ \text{V}$$

Example 1.8 A three-phase wye-connected source is connected to the wye con-nected load as shown in Fig. 1.24. The line voltage and the frequency of the source are 420 V and 50 Hz, respectively. For *abc* phase sequence, find the (i) line cur-rents, and (ii) power factor.

Solution

The phase voltages are,

$$V_{an} = \frac{V_L}{\sqrt{3}}\underline{|-30°} = \frac{420}{\sqrt{3}}\underline{|-30°} = 242.49\underline{|-30°}\ \text{V}$$
$$V_{bn} = 242.49\underline{|-150°}\ \text{V}$$
$$V_{cn} = 242.49\underline{|90°}\ \text{V}$$
$$Z_P = 10 + j15 = 18.03\underline{|56.31°}\ \Omega$$

Fig. 1.24 Wye-connected source and load

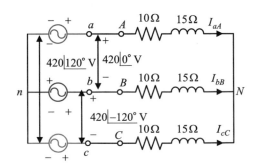

(i) The values of the line currents are,

$$I_{L1} = I_{aA} = \frac{242.49|-30°}{18.03|56.31°} = 13.45|-86.31°\ A$$

$$I_{L2} = I_{bB} = \frac{242.49|-150°}{18.03|56.31°} = 13.45|-206.31°\ A$$

$$I_{L3} = I_{cC} = \frac{242.49|90°}{18.03|56.31°} = 13.45|33.69°\ A$$

(ii) The value of the power factor is,

$$pf = \cos\phi = \frac{R}{Z_P} = \frac{10}{18} = 0.6\,\text{lag}$$

Practice problem 1.7
Figure 1.25 shows a three-phase wye-connected generator. For *abc* phase sequence, find the (i) values of the angles of the phase voltages, and (ii) magnitude of the line voltage.

Practice problem 1.8
Figure 1.26 shows a balanced wye-connected load is driven by a balanced three-phase wye-connected source. The source line voltage and the frequency are 300 V and 50 Hz, respectively. For *abc* phase sequence, find the line currents.

Fig. 1.25 Wye-connected
source

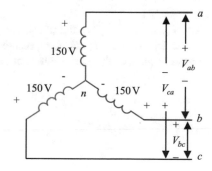

Fig. 1.26 Wye-connected
source and load

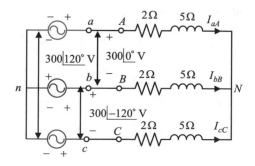

(a)   **(b)**   **(c)**

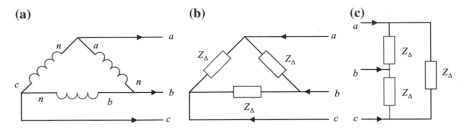

Fig. 1.27 Different delta configurations. **a** Generator. **b** Load. **c** Load

1.12   Delta or Mesh Connection

In delta system, the windings (of coils) are placed in a delta configuration, and this connection has also three output terminals as shown in Fig. 1.27. The delta-connection can be formed by connecting point n of a-n winding to point b of b-n winding, the point n of b-n winding to point c of c-n winding, and point n of c-n winding to point a of a-n winding. In this connection, the points to remember are (i) phase voltage is equal to the line voltage, and (ii) line current is equal to $\sqrt{3}$ times of the phase current.

1.13   Voltage and Current Relations for Delta-Connection

The I_{ab}, I_{bc} and I_{ca} are the phase currents in the delta-connected load as shown in Fig. 1.28. Also, I_a, I_b and I_c represents the line currents in this connection. For abc phase sequence, the phase currents are,

$$I_{ab} = I_P \underline{|0°}$$ (1.74)

$$I_{bc} = I_P \underline{|-120°}$$ (1.75)

$$I_{ca} = I_P \underline{|-240°}$$ (1.76)

Fig. 1.28 Delta connected load

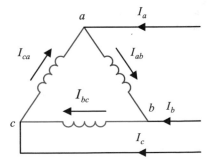

Applying KCL at nodes a, b and c in Fig. 1.28 yields,

$$I_a = I_{ab} - I_{ca} \qquad (1.77)$$

Substituting Eqs. (1.74) and (1.76) into Eq. (1.77) yields,

$$I_a = I_P\underline{|0°} - I_P\underline{|-120°} = \sqrt{3}\,I_P\underline{|-30°} \qquad (1.78)$$

$$I_b = I_{bc} - I_{ab} \qquad (1.79)$$

Substituting Eqs. (1.74) and (1.75) into Eq. (1.79) yields,

$$I_b = I_P\underline{|-120°} - I_P\underline{|0°} = \sqrt{3}\,I_P\underline{|-150°} \qquad (1.80)$$

$$I_c = I_{ca} - I_{bc} \qquad (1.81)$$

Substituting Eqs. (1.75) and (1.76) into Eq. (1.81) yields,

$$I_c = I_P\underline{|-240°} - I_P\underline{|-120°} = \sqrt{3}\,I_P\underline{|90°} \qquad (1.82)$$

Figure 1.29 shows a phasor diagram where the phase current I_{ab} is arbitrarily chosen as reference. From Eqs. (1.78), (1.80) and (1.82), it is found that the magnitude of the line current is $\sqrt{3}$ times greater than the magnitude of the phase current. The general relationship between the magnitude of line current and the phase current is,

$$I_L = \sqrt{3}\,I_P \qquad (1.83)$$

From Fig. 1.29, it is seen that the phase voltage is equal to the line voltage i.e., the relationship is,

$$V_L = V_P \qquad (1.84)$$

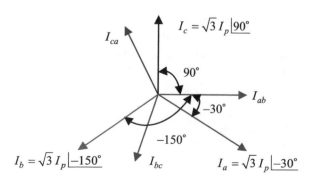

Fig. 1.29 Phasor diagram with currents

Fig. 1.30 Delta connection
load

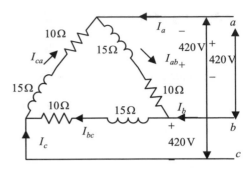

Example 1.9 Figure 1.30 shows a balanced three-phase delta load having per phase
impedance of $10+j15$ Ω. The load is connected with the 420 V, three-phase,
delta-connected supply. For abc phase sequence, calculate the (i) phase currents,
and (ii) magnitude of the line current.

Solution

The value of the per phase impedance is,

$$Z_P = 10+j15 = 18.03\underline{|56.31°}\ \Omega$$

(i) The values of the phase currents are,

$$I_{ab} = \frac{V_{ab}}{Z_P} = \frac{420\underline{|0°}}{18.03\underline{|56.31°}} = 23.29\underline{|-56.31°}\ \text{A}$$

$$I_{bc} = \frac{V_{bc}}{Z_P} = \frac{420\underline{|-120°}}{18.03\underline{|56.31°}} = 23.29\underline{|-176.31°}\ \text{A}$$

$$I_{ca} = \frac{V_{ca}}{Z_P} = \frac{420\underline{|120°}}{18.03\underline{|56.31°}} = 23.29\underline{|63.69°}\ \text{A}$$

(ii) The magnitude of line current is,

$$I_L = \sqrt{3}I_P = 1.732 \times 23.29 = 40.33\ \text{A}$$

Practice problem 1.9

Figure 1.31 shows a balanced three-phase delta-connected load. For abc phase
sequence, determine the (i) value of the phase voltages with angles, (ii) phase
currents, and (iii) magnitude of the line current.

Fig. 1.31 Delta connection
load

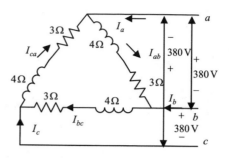

1.14   Three-Phase Power Calculation

Consider instantaneous values of voltages and currents of Y-connected load for calculating the total power of the three-phase system. The instantaneous voltages are,

$$v_{an} = V_m \sin \omega t \tag{1.85}$$

$$v_{bn} = V_m \sin(\omega t - 120°) \tag{1.86}$$

$$v_{cn} = V_m \sin(\omega t - 240°) \tag{1.87}$$

Again, consider that the phase impedance of the Y-connected load is $Z_Y = Z\angle\theta$. Then the phase currents lag behind their corresponding phase voltages by θ. Therefore, the phase currents can be expressed as,

$$i_a = \frac{v_{an}}{Z\underline{|\theta}} = \frac{V_m \sin \omega t}{Z\underline{|\theta}} = I_m \sin(\omega t - \theta) \tag{1.88}$$

$$i_b = I_m \sin(\omega t - \theta - 120°) \tag{1.89}$$

$$i_c = I_m \sin(\omega t - \theta - 240°) \tag{1.90}$$

Instantaneous power for phase a can be expressed as,

$$p_a = \frac{1}{T} \int_0^T v_{an} \times i_a \, dt \tag{1.91}$$

Substituting Eqs. (1.85) and (1.88) into Eq. (1.91) yields,

$$p_a = \frac{V_m I_m}{T} \int_0^T \sin \omega t \times \sin(\omega t - \theta) \, dt \tag{1.92}$$

$$p_a = \frac{V_m I_m}{2T} \int_0^T 2 \sin \omega t \times \sin(\omega t - \theta) \, dt \tag{1.93}$$

$$p_a = \frac{V_m I_m}{2T} \int_0^T [\cos \theta - \cos(2\omega t - \theta)] \, dt \tag{1.94}$$

$$p_a = \frac{V_m I_m}{2T} \times \cos \theta \times T - 0 \tag{1.95}$$

$$p_a = \frac{V_m I_m}{\sqrt{2} \times \sqrt{2}} \cos \theta = V_P I_P \cos \theta \tag{1.96}$$

Similarly, the expressions of the instantaneous power for phases b and c can be written as,

$$p_b = V_p I_p \cos \theta \tag{1.97}$$

$$p_c = V_p I_p \cos \theta \tag{1.98}$$

Therefore, the average three-phase power, P can be calculated as,

$$P = p_a + p_b + p_c = 3 V_p I_p \cos \theta \tag{1.99}$$

Similarly, the expression of reactive power can be expressed as,

$$Q = 3 V_p I_p \sin \theta \tag{1.100}$$

where V_p and I_p are the rms values of phase voltage and phase current, respectively.
Y-connection: Substituting Eqs. (1.72) and (1.73) into Eq. (1.99) yields,

$$P_Y = 3 \times \frac{V_L}{\sqrt{3}} \times I_L \times \cos \theta \tag{1.101}$$

$$P_Y = \sqrt{3} \, V_L I_L \cos \theta \tag{1.102}$$

Substituting Eqs. (1.72) and (1.73) into Eq. (1.100) yields,

$$Q_Y = 3 \times \frac{V_L}{\sqrt{3}} \times I_L \times \sin \theta \tag{1.103}$$

$$Q_Y = \sqrt{3} \, V_L I_L \sin \theta \tag{1.104}$$

Delta connection: Again, substituting Eqs. (1.72) and (1.73) into Eq. (1.99) yields,

$$P_\Delta = 3 \times \frac{V_L}{\sqrt{3}} \times I_L \times \cos\theta \qquad (1.105)$$

$$P_\Delta = \sqrt{3}\, V_L I_L \cos\theta \qquad (1.106)$$

Substituting Eqs. (1.72) and (1.73) into Eq. (1.100) yields,

$$Q_\Delta = 3 \times \frac{V_L}{\sqrt{3}} \times I_L \times \sin\theta \qquad (1.107)$$

$$Q_\Delta = \sqrt{3}\, V_L I_L \sin\theta \qquad (1.108)$$

Example 1.10 Figure 1.32 shows a balanced Y-Y system where $Z_y = 3 + j4\,\Omega$ and $V_{AN} = 150\underline{|0°}$ V. For *ABC* phase sequence, calculate the (i) line currents, (ii) power factor, (iii) power supplied to each phase, and (iv) total power supplied to the load.

Solution

(i) The source voltages are,

$$V_{AN} = 150\underline{|0°}\ \text{V}$$
$$V_{BN} = 150\underline{|-120°}\ \text{V}$$
$$V_{CN} = 150\underline{|120°}\ \text{V}$$

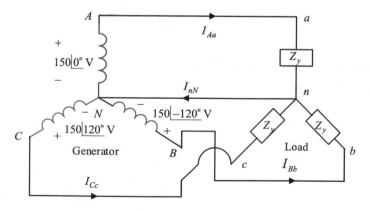

Fig. 1.32 Balanced Y-Y system

The line currents are,

$$I_{Aa} = \frac{V_{AN}}{Z_y} = \frac{150\lfloor 0°}{5\lfloor 53.13°} = 30\lfloor -53.13° \text{ A}$$

$$I_{Bb} = \frac{V_{BN}}{Z_y} = \frac{150\lfloor -120°}{5\lfloor 53.13°} = 30\lfloor -173.13° \text{ A}$$

$$I_{Cc} = \frac{V_{CN}}{Z_y} = \frac{150\lfloor 120°}{5\lfloor 53.13°} = 30\lfloor 66.87° \text{ A}$$

(ii) The power factor is,

$$pf = \cos\theta = \frac{R}{Z} = \frac{3}{5} = 0.6$$

The power supplied to each phase is,

$$P = VI\cos\theta = 150 \times 30 \times 0.6 = 2.7\,\text{kW}$$

The total power supplied is,

$$P_t = 3P = 3 \times 2.7 = 8.1\,\text{kW}$$

Example 1.11 A balanced three-phase star-connected load is shown in Fig. 1.33. Find the (i) line current, (ii) power factor, and (iii) total power supplied.

Solution

(i) The value of the phase voltage is,

$$V_p = \frac{V_L}{\sqrt{3}} = \frac{400}{\sqrt{3}} = 230.94 \text{ V}$$

Fig. 1.33 A delta connected load

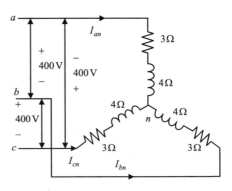

The value of the impedance is,

$$Z_y = \sqrt{3^2 + 4^2} = 5\,\Omega$$

The value of the phase current is,

$$I_p = \frac{V_p}{Z_y} = \frac{230.94}{5} = 46.19\,\Omega$$

The value of the line current is,

$$I_L = I_p = 46.19\,\text{A}$$

(ii)   Power factor is,

$$pf = \cos\theta = \frac{R}{Z} = \frac{3}{5} = 0.6$$

(iii)   The total power supplied,

$$P = \sqrt{3}V_L I_L \cos\theta = \sqrt{3} \times 400 \times 46.19 \times 0.6 = 19.2\,\text{kW}$$

Practice problem 1.10
Figure 1.34 shows a balanced $\Delta - Y$ system where $Z_y = 4 - j6\,\Omega$. For ABC phase sequence, find the (i) line currents, (ii) power factor, (iii) phase power, and (iv) total power supplied to the load.

Practice problem 1.11
Figure 1.35 shows a balanced three-phase star-connected load. Determine the (i) line currents, (ii) power factor, and (iii) total power supplied to the load.

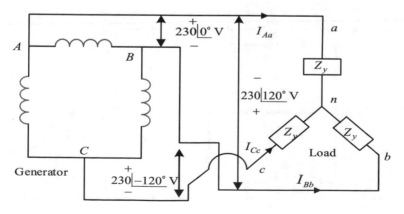

Fig. 1.34 A delta-wye system

Fig. 1.35 Star connected
load

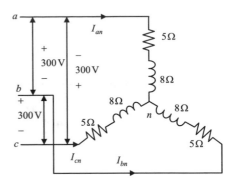

1.15 Measurement of Three-Phase Power

The instrument wattmeter is used for measuring the power of any power-line with
load. The wattmeter has two coils, namely, current coil that is connected in series
with the line, and the voltage or pressure coil, which is connected across the line.

A single-phase wattmeter can measure the total power of the balanced
three-phase system. Generally, the two-wattmeter method is used to measure the
three-phase power of the circuit. The connection diagram of two-wattmeter method
is shown in Fig. 1.36. Total instantaneous power is given by,

$$p = v_{an} i_a + v_{bn} i_b + v_{cn} i_c \qquad (1.109)$$

Here, v_{an}, v_{bn} and v_{cn} are the instantaneous voltages across the loads Z_1, Z_2 and
Z_3, respectively, and i_a, i_b and i_c are the instantaneous line (phase) currents. In the
absence of neutral connection, the total current at the neutral point is,

$$0 = i_a + i_b + i_c \qquad (1.110)$$

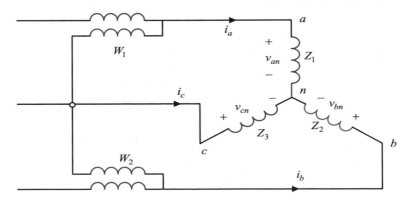

Fig. 1.36 Two-wattmeter method

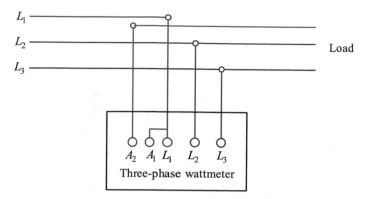

Fig. 1.37 Connection diagram of a three-phase wattmeter

$$i_b = -(i_a + i_c) \tag{1.111}$$

Substituting Eq. (1.111) into Eq. (1.109) yields,

$$p = v_{an}i_a - v_{bn}(i_a + i_c) + v_{cn}i_c \tag{1.112}$$

$$p = (v_{an} - v_{bn})i_a + (v_{cn} - v_{bn})i_c \tag{1.113}$$

A three-phase wattmeter is also used to measure the power of a three-phase line and this connection diagram is shown in Fig. 1.37.

1.16   Power Factor Measurement

Figure 1.38 shows the balanced Y-connected load and two wattmeters are connected with this load to measure the total power. Consider V_{an}, V_{bn} and V_{cn} are the rms values of the phase voltages across the loads respectively, and I_a, I_b and I_c are the rms values of the line currents respectively. Again, consider that the currents lag the corresponding phase voltages by an angle of θ. The average power recorded by the first wattmeter, W_1 is

$$P_1 = V_{ac}I_a \cos\theta_{ac} \tag{1.114}$$

The average power recorded by the second wattmeter, W_2 is,

$$P_2 = V_{bc}I_b \cos\theta_{bc} \tag{1.115}$$

where θ_{ac} and θ_{bc} are the phase angles between the respective phases. The magnitude of those phase angles can be determined with the help of phasor diagram as shown in Fig. 1.39. These values are,

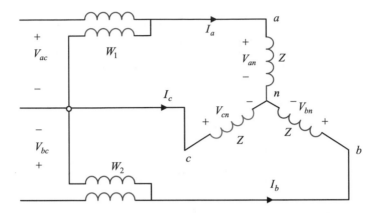

Fig. 1.38 Balanced wye-connected load with wattmeter

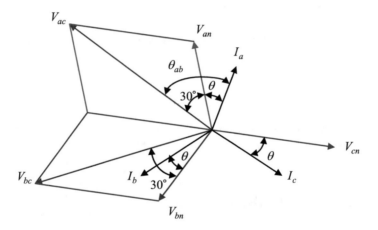

Fig. 1.39 Phasor diagram

$$\theta_{ac} = 30° + \theta \qquad (1.116)$$

$$\theta_{bc} = 30° - \theta \qquad (1.117)$$

The following relation can be written in case of Y-connected balanced load:

$$I_a = I_b = I_c = I_L \qquad (1.118)$$

$$V_{ab} = V_{bc} = V_{ca} = V_L \qquad (1.119)$$

Substituting related parameters into Eqs. (1.114) and (1.115) yields,

$$P_1 = V_L I_L \cos(30° + \theta) \tag{1.120}$$

$$P_2 = V_L I_L \cos(30° - \theta) \tag{1.121}$$

The sum of the wattmeter readings is,

$$P = P_1 + P_2 \tag{1.122}$$

Substituting Eqs. (1.120) and (1.121) into Eq. (1.122) yields,

$$P = V_L I_L \cos(30° + \theta) + V_L I_L \cos(30° - \theta) \tag{1.123}$$

$$P = 2V_L I_L \cos 30° \cos \theta \tag{1.124}$$

$$P_1 + P_2 = \sqrt{3} V_L I_L \cos \theta \tag{1.125}$$

Similarly, the difference of the wattmeter readings can be written as,

$$P_2 - P_1 = V_L I_L \sin \theta \tag{1.126}$$

Dividing Eq. (1.126) by Eq. (1.125) yields,

$$\frac{1}{\sqrt{3}} \tan \theta = \frac{P_2 - P_1}{P_2 + P_1} \tag{1.127}$$

$$\tan \theta = \sqrt{3} \frac{P_2 - P_1}{P_2 + P_1} \tag{1.128}$$

Therefore, the power factor angle θ and the power factor can be determined using Eq. (1.128).

Example 1.12 The two-wattmeter method is used to measure the total power of a three-phase circuit and is found to be 120 kW. Calculate the second wattmeter reading if the power factor is 0.6.

Solution

The power factor angle can be determined as,

$$\theta = \cos^{-1} 0.6 = 53.13°$$
$$\tan \theta = \sqrt{3} \frac{P_2 - P_1}{P_2 + P_1}$$
$$\tan 53.13 = \sqrt{3} \frac{P_2 - P_1}{120}$$

The difference in power is,

$$P_2 - P_1 = 92.4$$

The sum of the power is,

$$P_2 + P_1 = 120$$

The second wattmeter reading can be determined as,

$$P_2 = 106.2 \, \text{kW}$$

Practice problem 1.12
The power factor of a three-phase, 420 V induction motor is 0.6. The input power is measured by the two-wattmeter method and found to be 35 kW. Find the reading on the second wattmeter and the line current.

1.17   Series Resonance

For series resonance, Fig. 1.40 shows the RLC series circuit. Here, the inductive and the capacitive reactance are the frequency dependent. If the frequency increases, the inductive reactance increases and capacitive reactance decreases. At resonance, the circuit power factor is unity. For a unity power factor, the net reactance in the series circuit is zero. Therefore, at resonance condition, the circuit impedance is equal to resistance.

At a certain frequency, the complex part of the total impedance is zero, i.e., the inductive reactance is equal to the capacitive reactance. This frequency is known as resonant frequency and it is represented by f_0. The equivalent impedance of the RLC series circuit is,

$$Z_t = R + j(X_L - X_C) \tag{1.129}$$

At series resonance, the complex part of Eq. (1.129) is equal to zero and it can be expressed as,

Fig. 1.40 An RLC series circuit

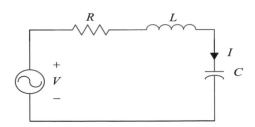

$$X_L - X_C = 0 \tag{1.130}$$

$$2\pi f_0 L = \frac{1}{2\pi f_0 C} \tag{1.131}$$

$$f_0 = \frac{1}{2\pi\sqrt{LC}} \tag{1.132}$$

$$\omega_0 = \frac{1}{\sqrt{LC}} \tag{1.133}$$

From Eq. (1.132) or (1.133), it is seen that the value of the resistance in the circuit has no effect on the resonant frequency.

The circuit with a resistor can dissipate more energy. It is not possible to build a circuit with pure inductor and capacitor. Practically, these elements have some resistances. The quality factor is defined as the ratio of voltage across the inductance or the capacitance to the voltage across the pure resistance. Alternatively, it is defined as the ratio of the reactive power delivered in either the inductor or the capacitor to the average power delivered to the resistor at resonance. The sharpness of the resonance in a resonant circuit can be identified by the quality factor or the Q_s-factor. According to the first and second definitions, the Q_s-factor can be expressed as,

$$Q_s = \frac{V_L}{V_R} \tag{1.134}$$

$$Q_s = \frac{I^2 X_L}{I^2 R} = \frac{X_L}{R} \tag{1.135}$$

$$Q_s = \frac{I^2 X_C}{I^2 R} = \frac{X_C}{R} \tag{1.136}$$

Equation (1.135) can be re-arranged as,

$$Q_s = \frac{2\pi f_0 L}{R} \tag{1.137}$$

Substitute Eq. (1.132) into Eq. (1.137) and the Q_s-factor becomes,

$$Q_s = \frac{2\pi L}{R} \times \frac{1}{2\pi\sqrt{LC}} = \frac{L}{R}\frac{1}{\sqrt{LC}} \tag{1.138}$$

$$Q_s = \frac{1}{R}\sqrt{\frac{L}{C}} \tag{1.139}$$

Equation (1.139) can also be derived from Eq. (1.136) as,

$$Q_s = \frac{1}{2\pi f_0 CR} \tag{1.140}$$

Substitute Eq. (1.132) into Eq. (1.140) and the Q_s-factor becomes,

$$Q_s = \frac{1}{2\pi CR} \times \frac{1}{\frac{1}{2\pi\sqrt{LC}}} \tag{1.141}$$

$$Q_s = \frac{1}{2\pi CR} \times 2\pi\sqrt{LC} \tag{1.142}$$

$$Q_s = \frac{1}{R}\sqrt{\frac{L}{C}}$$

1.18 Parallel Resonance

A RLC parallel circuit where the inductor contains a small resistance is shown in Fig. 1.41. This circuit is said to be in resonance when the power factor is unity, i.e., the reactive component of the current in the inductive branch is equal to the reactive component of the current in the capacitive branch. The phasor diagrams for different currents are shown in Fig. 1.42. The reactive component of the current in the inductive branch is,

$$I_{RL} = I_L \sin \phi \tag{1.143}$$

The reactive component of the current for the capacitive branch is,

$$I_{RC} = I_C \tag{1.144}$$

Fig. 1.41 A RLC parallel ac circuit

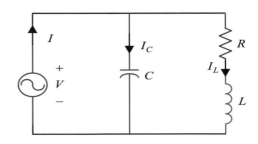

Fig. 1.42 Phasor diagram for currents

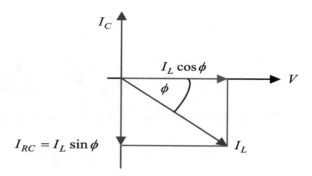

The current in the capacitive reactance is,

$$I_C = \frac{V}{X_C} \tag{1.145}$$

The current in the inductive branch is,

$$I_L = \frac{V}{Z_L} \tag{1.146}$$

$$\sin \phi = \frac{X_L}{Z_L} \tag{1.147}$$

According to the definition of parallel resonance, the following equation can be written as,

$$I_C = I_L \sin \phi \tag{1.148}$$

Substituting Eqs. (1.145), (1.146) and (1.147) into Eq. (1.143) provides,

$$\frac{V}{X_C} = \frac{V}{Z_L} \times \frac{X_L}{Z_L} \tag{1.149}$$

$$Z_L^2 = X_L X_C \tag{1.150}$$

$$Z_L^2 = \frac{L}{C} \tag{1.151}$$

Equation (1.150) can also be derived in an alternative way. The parallel circuit would be in resonance when the j-component of the total admittance is zero. The total admittance of the circuit as shown in the Fig. 1.41 is,

$$Y_t = Y_L + Y_C = \frac{1}{R + jX_L} + \frac{1}{-jX_C} \tag{1.152}$$

$$Y_t = \frac{R - jX_L}{R^2 + X_L^2} - \frac{1}{jX_C} \tag{1.153}$$

$$Y_t = \frac{R}{R^2 + X_L^2} + j\left(\frac{1}{X_C} - \frac{X_L}{R^2 + X_L^2}\right) \tag{1.154}$$

Consider the j-component of Eq. (1.154) is equal to zero which becomes,

$$\frac{1}{X_C} - \frac{X_L}{R^2 + X_L^2} = 0 \tag{1.155}$$

$$X_L X_C = R^2 + X_L^2 = Z_L^2 \tag{1.156}$$

Substituting $X_L = \omega L$ and $X_C = \frac{1}{\omega C}$ into Eq. (1.156) yields,

$$\frac{\omega L}{\omega C} = R^2 + X_L^2 = Z_L^2 \tag{1.157}$$

$$\frac{L}{C} = R^2 + (2\pi f_0 L)^2 \tag{1.158}$$

$$(2\pi f_0 L)^2 = \frac{L}{C} - R^2 \tag{1.159}$$

$$f_0 = \frac{1}{2\pi} \sqrt{\frac{1}{LC} - \frac{R^2}{L^2}} \tag{1.160}$$

From Eq. (1.154), the impedance at resonance can be derived as,

$$Y_t = \frac{1}{Z_r} = \frac{R}{R^2 + X_L^2} + 0 \tag{1.161}$$

Equation (1.161) can be modified as,

$$\frac{1}{Z_r} = \frac{R}{Z_L^2} \tag{1.162}$$

Substituting Eq. (1.157) into Eq. (1.162) yields,

$$\frac{1}{Z_r} = \frac{R}{\frac{L}{C}} \tag{1.163}$$

$$Z_r = \frac{L}{RC} \tag{1.164}$$

From Eq. (1.164), it is observed that the impedance at resonance condition is equal to the ratio of the inductance to the product resistance and capacitance.

Exercise Problems

1.1 The excitation voltage of the circuit as shown in Fig. P1.1 is $v(t) = 20\sin(\omega t - 45°)$ V and the value of the impedance is $Z = 3\underline{|-30°}$ Ω. Find the current in time domain, and the instantaneous power.

1.2 The current and the impedance of the circuit as shown in Fig. P1.2 are $i(t) = 2\sin(314t + 35°)$ A and $Z = 5\underline{|45°}$ Ω respectively. Find the voltage $v(t)$ and the instantaneous power.

1.3 The current of $5\underline{|38°}$ A is flowing through the impedance of $5\underline{|34°}$ Ω. Determine the average power absorbed by the impedance.

1.4 The expression of current through the 4 Ω resistance is given by $i(t) = 2\sin 3t$ A. Calculate the average power absorbed by the resistance.

1.5 The expression of current in the $2 + j3$ Ω impedance is given by $i(t) = 2\sin 3t - 5\sin 4t$ A. Find the average power absorbed by the impedance.

1.6 The current of $i(t) = 3\sin t - 2\sin 2t + 6\sin 3t$ A flows through the impedance of $4 - j10$ Ω. Calculate the average power absorbed by the impedance.

1.7 Determine the average power of each element of the circuit as shown in Fig. P1.3. The value of the voltage excitation is $v(t) = 5\sin(314t + 36°)$ V.

1.8 In the circuit as shown in Fig. P1.4, find the total average power supplied and the power absorbed by the 3 Ω and 4 Ω resistors respectively.

1.9 Determine the power supplied by the voltage source and the power absorbed by the resistive elements of the circuit as shown in Fig. P1.5.

1.10 Determine the total average power supplied by the source and the average power absorbed by the 2 Ω resistor of the circuit as shown in Fig. P1.6.

Fig. P1.1 A simple series ac circuit

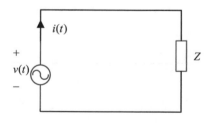

Fig. P1.2 A simple series ac circuit

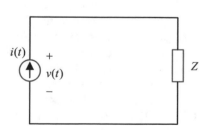

Fig. P1.3 A simple RC series
circuit with voltage source

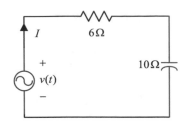

Fig. P1.4 A simple parallel
ac circuit with current source

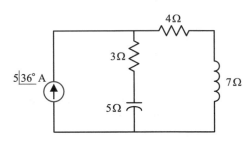

Fig. P1.5 A simple parallel
ac circuit

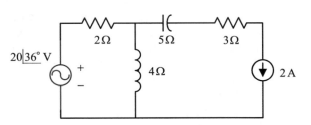

Fig. P1.6 A parallel ac
circuit with voltage source

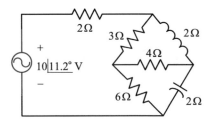

1.11 Figure P1.7 shows a parallel ac circuit. Calculate the (i) source current, (ii) total complex power, (iii) total real power, and (iv) total reactive power.

1.12 Find the source current, total complex power and total real power of the circuit as shown in Fig. P1.8.

1.13 Figure P1.9 shows an induction motor that is connected across the 220 V rms, 50 Hz source. A capacitor is connected across the induction motor to raise the power factor to 0.95 lagging. Determine the value of the capacitor. Also determine the value of the capacitor if it is used to convert the power factor to 0.85 leading.

1.14 A 15 kW, wye-connected, three-phase induction motor with a lagging power factor of 0.85 is connected to an 220 V rms, 50 Hz source. A capacitor is

Fig. P1.7 A simple parallel
ac circuit

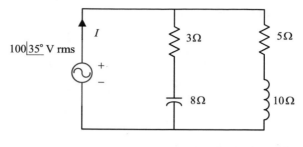

Fig. P1.8 A simple series ac
circuit

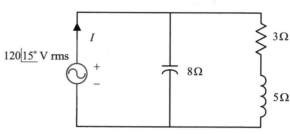

Fig. P1.9 A parallel load
with capacitor

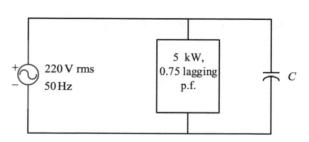

connected across the motor to raise the power factor to 0.95 lagging. Find the
value of the capacitor.

1.15 A 12 kW, delta-connected, three-phase induction motor with a lagging power
factor of 0.80 is connected to the 120 V rms, 50 Hz source. A capacitor is
connected across the induction motor to improve the power factor to 0.95
leading. Calculate the value of the capacitor.

1.16 Figure P1.10 shows the circuit where the load is connected to the source
through the transmission line. Find the line current, the source voltage and
the source power factor.

1.17 In the circuit shown in Fig. P1.11, determine the value of the source current.

1.18 A three-phase Y-connected generator is shown in the Fig. P1.12. The
magnitude of the phase voltage is 130 V rms. For *abc* phase sequence, write
down the phase voltages with angles, and calculate the line voltage.

1.19 A balanced three-phase Y-connected load is shown in the Fig. P1.13. The
magnitude of the line-to-line voltage is 380 V rms. The resistance of each
phase of the load is 15 Ω. For *abc* phase sequence, write down the phase
voltages, and find the phase currents in the load.

Fig. P1.10 A series load

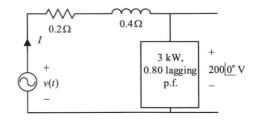

Fig. P1.11 A parallel load

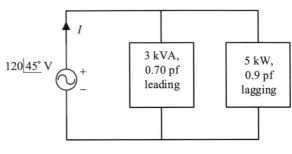

Fig. P1.12 A wye-connected
generator

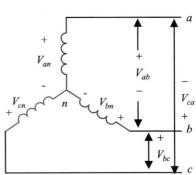

Fig. P1.13 A wye-connected
load

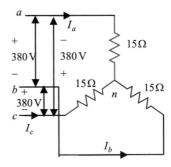

1.20   A wye-connected load with three coils, each having a resistance of $6\,\Omega$ and
       an inductive reactance of $8\,\Omega$ is shown in the Fig. P1.14. For *abc* phase
       sequence and delta-connected source, find (i) the line currents with phase
       angles, (ii) the neutral current, and (iii) power factor.

1.21   An unbalanced three-phase Y-connected load is shown in the Fig. P1.15. The
       magnitude of the line voltage is 200 V rms. For *abc* phase sequence and
       delta-connected source, calculate (i) the line currents with angles, and (ii) the
       neutral current.

1.22   A balanced three-phase Y-connected load is driven by Y-connected source as
       shown in Fig. P1.16. Each phase of the load is having a resistance of $3\,\Omega$ and
       a capacitive reactance of $6\,\Omega$. For *abc* phase sequence, determine the
       (i) phase voltages, (ii) line currents with phase angles, and (ii) power factor.

1.23   A three-phase Y-connected load is driven by Y-connected source as shown in
       the Fig. P1.17. The magnitude of the line-to-line voltage is 200 V rms and the

Fig. P1.14 A wye-connected
load

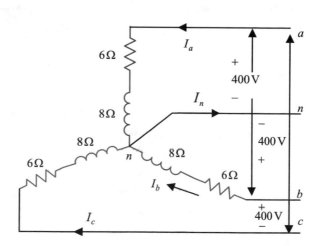

Fig. P1.15 A wye-connected
load

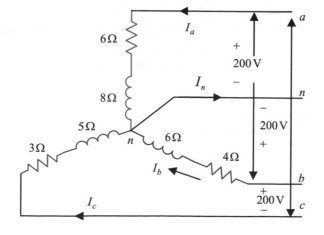

Fig. P1.16 A wye-connected load

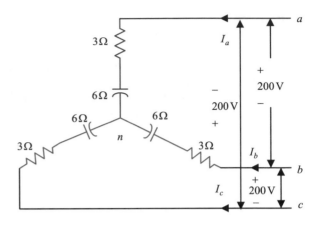

Fig. P1.17 A wye-connected generator

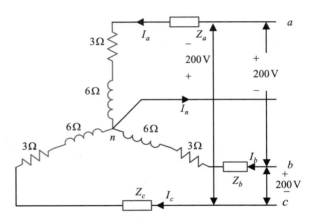

line impedances are $Z_a = Z_b = Z_c = 2 + j3\,\Omega$. For *abc* phase sequence, calculate the (i) line currents with phase angles, and (ii) power factor of the load.

1.24 A balanced three-phase, Δ-connected load is shown in the Fig. P1.18. The per phase impedance of the load is $4 + j6\,\Omega$. If 415 V, three-phase supply is connected to this load, find the magnitude of (i) the phase current, and (ii) the line current.

Fig. P1.18 A delta-connected generator

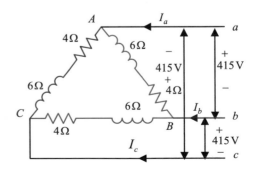

References

1. C.K. Alexander, M.N.O. Sadiku, *Fundamentals of Electric Circuits*, 5th edn (McGraw-Hill Higher Education, 12 March 2014)
2. M.A. Salam, *Electrical Circuits*, 1st edn. (Alpha Science International Ltd., Oxford, 2007)
3. J.D. Irwin, R.M. Nelms, *Basic Engineering Circuit Analysis*, 10th edn (Wiley, 2011)

Chapter 2
Transformer: Principles and Practices

2.1 Introduction

There are many devices such as three-phase ac generator, transformer etc., which are usually used in a power station to generate and supply electrical power to a power system network. In the power station, the three-phase ac generator generates a three-phase alternating voltage in the range between 11 and 20 kV. The magnitude of the generated voltage is increased to 120 kV or more by means of a power transformer. This higher magnitude of voltage is then transmitted to the grid substation by a three-phase transmission lines. A lower line voltage of 415 V is achieved by stepping down either from the 11 or 33 kV lines by a distribution transformer. In these cases, a three-phase transformer is used in either to step-up or step-down the voltage. Since a transformer plays a vital role in feeding an electrical network with the required voltage, it becomes an important requirement of a power system engineer to understand the fundamental details about a transformer along with its analytical behavior in the circuit domain. This chapter is dedicated towards this goal. On the onset of this discussion it is worth mentioning that a transformer, irrespective of its type, contains the following characteristics (i) it has no moving parts, (ii) no electrical connection between the primary and secondary windings, (iii) windings are magnetically coupled, (iv) rugged and durable in construction, (v) efficiency is very high i.e., more than 95 %, and (vi) frequency is unchanged.

2.2 Working Principle of Transformer

Figure 2.1 shows a schematic diagram of a single-phase transformer. There are two types of windings in a single-phase transformer. These are called primary and secondary windings or coils. The primary winding is connected to the alternating

© Springer Science+Business Media Singapore 2016
Md.A. Salam and Q.M. Rahman, *Power Systems Grounding*,
Power Systems, DOI 10.1007/978-981-10-0446-9_2

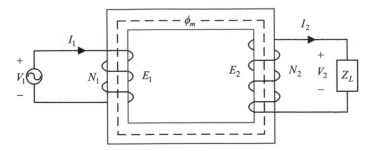

Fig. 2.1 Single-phase transformer

voltage source and the secondary winding is connected to the load. The primary and secondary winding parameters are represented by the suffix p or 1 and s or 2, respectively. A sinusoidal current flows in the primary winding when it is connected to an alternating voltage source. This current establish a flux φ which moves from the primary winding to the secondary winding through low reluctance magnetic core.

About 95 % of this flux moves from the primary to the secondary through the low reluctance path of the magnetic core and this flux is linked by the both windings and a small percent of this flux links to the primary winding. According to the Faradays laws of electromagnetic induction, a voltage will be induced across the secondary winding as well as in the primary winding. Due to this voltage, a current will flow through the load if it is connected with the secondary winding. Hence, the primary voltage is transferred to the secondary winding without a change in frequency.

2.3 Flux in a Transformer

The current in the primary winding establishes a flux. The flux that moves from primary to secondary and links both the windings is called the mutual flux and its maximum value is represented by ϕ_m.

Flux which links only the primary winding and completes the magnetic path through the surrounding air is known primary leakage flux. The primary leakage flux is denoted by ϕ_{1l}. Similarly, secondary leakage flux is that flux which links only the secondary winding and completes the magnetic path through the surrounding air. The secondary leakage flux is denoted by ϕ_{2l}. Mutual and leakage fluxes are shown in Fig. 2.2.

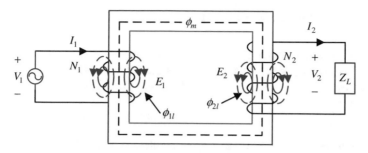

Fig. 2.2 Mutual and leakage fluxes

2.4 Ideal Transformer

An ideal transformer is one which does not supply any energy to the load i.e., the secondary winding is open circuited. The main points of an ideal transformer are (i) no winding resistance, (ii) no leakage flux and leakage inductance, (iii) self-inductance and mutual inductance are zero, (iv) no losses due to resistance, inductance, hysteresis or eddy current and (v) coefficient of coupling is unity.

Figure 2.3a shows an ideal transformer where the secondary winding is left open. A small magnetizing current I_m will flow in the primary winding when it is connected to the alternating voltage source, V_1. This magnetizing current lags behind the supply voltage, V_1 by 90° and produces the flux φ, which induces the primary and secondary emfs. These emfs lag behind the flux, φ by 90°. The magnitude of primary induced emf E_1 and supply voltage V_1 is the same, but are 180° out of phase as shown in Fig. 2.3b.

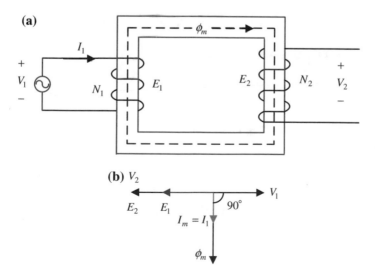

Fig. 2.3 Ideal transformer and phasor diagram

2.5 E.M.F. Equation of Transformer

The primary winding draws a current when it is connected to an alternating voltage source. This primary sinusoidal current produces a sinusoidal flux φ that can be expressed as,

$$\phi = \phi_m \sin \omega t \tag{2.1}$$

Instantaneous emf induced in the primary winding is,

$$e_1 = -N_1 \frac{d\phi}{dt} \tag{2.2}$$

Similarly, instantaneous emf induced in the secondary winding is,

$$e_2 = -N_2 \frac{d\phi}{dt} \tag{2.3}$$

Substituting Eq. (2.1) into the Eq. (2.2) yields,

$$e_1 = -N_1 \frac{d}{dt} (\phi_m \sin \omega t) \tag{2.4}$$

$$e_1 = -N_1 \omega \phi_m \cos \omega t \tag{2.5}$$

$$e_1 = N_1 \omega \phi_m \sin(\omega t - 90°) \tag{2.6}$$

The maximum value of e_1 is,

$$E_{m1} = N_1 \omega \phi_m \tag{2.7}$$

The rms value of the primary emf is,

$$E_1 = \frac{E_{m1}}{\sqrt{2}} \tag{2.8}$$

Substituting Eq. (2.7) into Eq. (2.8) yields,

$$E_1 = \frac{N_1 2\pi f \phi_m}{\sqrt{2}} \tag{2.9}$$

$$E_1 = 4.44 f \phi_m N_1 \tag{2.10}$$

Similarly, the expression of the secondary emf is,

$$E_2 = 4.44 f \phi_m N_2 \qquad (2.11)$$

The primary and secondary voltages can be determined from Eqs. (2.10) and (2.11) if other parameters are known.

2.6   Turns Ratio of Transformer

Turns ratio is an important parameter for drawing an equivalent circuit of a transformer. The turns ratio is used to identify the step-up and step-down transformers. According to Faraday's laws, the induced emf in the primary (e_1) and the secondary (e_2) windings are,

$$e_1 = -N_1 \frac{d\phi}{dt} \qquad (2.12)$$

$$e_2 = -N_2 \frac{d\phi}{dt} \qquad (2.13)$$

Dividing Eq. (2.12) by Eq. (2.13) yields,

$$\frac{e_1}{e_2} = \frac{N_1}{N_2} \qquad (2.14)$$

$$\frac{e_1}{e_2} = \frac{N_1}{N_2} = a \qquad (2.15)$$

Similarly, dividing Eq. (2.10) by (2.11) yields,

$$\frac{E_1}{E_2} = \frac{N_1}{N_2} = a \qquad (2.16)$$

where a is the turns ratio of a transformer. In case of $N_2 > N_1$, the transformer is called a step-up transformer. Whereas for $N_1 > N_2$, the transformer is called a step-down transformer. The losses are zero in an ideal transformer. In this case, the input power of the transformer is equal to its output power and this yields,

$$V_1 I_1 = V_2 I_2 \qquad (2.17)$$

Equation (2.17) can be rearranged as,

$$\frac{V_1}{V_2} = \frac{I_2}{I_1} = a \qquad (2.18)$$

The ratio of primary current to the secondary current is,

$$\frac{I_1}{I_2} = \frac{1}{a} \tag{2.19}$$

Again, the magnetomotive force produced by the primary current will be equal to the magnetomotive force produced by the secondary current and it can be expressed as,

$$\Im = \Im_1 - \Im_2 = 0 \tag{2.20}$$

$$N_1 I_1 = N_2 I_2 \tag{2.21}$$

$$\frac{I_1}{I_2} = \frac{N_2}{N_1} = \frac{1}{a} \tag{2.22}$$

From Eq. (2.22), it is concluded that the ratio of primary to secondary current is inversely proportional to the turns ratio of the transformer.

The input and output power of an ideal transformer is,

$$P_{in} = V_1 I_1 \cos \phi_1 \tag{2.23}$$

$$P_{out} = V_2 I_2 \cos \phi_2 \tag{2.24}$$

For an ideal condition, the angle ϕ_1 is equal to the angle ϕ_2 and the output power can be re-arranged as,

$$P_{out} = \frac{V_1}{a} a I_1 \cos \phi_1 \tag{2.25}$$

$$P_{out} = V_1 I_1 \cos \phi_1 = P_{in} \tag{2.26}$$

From Eq. (2.26), it is seen that the input and the output power are the same in case of an ideal transformer. Similarly, the input and output reactive powers are,

$$Q_{out} = V_2 I_2 \sin \phi_2 = V_1 I_1 \sin \phi_1 = Q_{in} \tag{2.27}$$

From Eqs. (2.26) and (2.27), the input and output power and reactive power can be calculated if other parameters are given.

Example 2.1 The number of turns in the secondary coil of a 22 kVA, 2200 V/220 V single-phase transformer is 50. Find the (i) number of primary turns, (ii) primary full load current, and (iii) secondary full load current. Neglect all kinds of losses in the transformer.

Solution

The value of the turns ratio is,

$$a = \frac{V_1}{V_2} = \frac{2200}{220} = 10$$

(i) The value of the primary turns can be determined as,

$$a = \frac{N_1}{N_2}$$
$$N_1 = aN_2 = 10 \times 50 = 500$$

(ii) The value of the primary full load current is,

$$I_1 = \frac{22 \times 10^3}{2200} = 10\,\text{A}$$

(iii) The value of the secondary full load current is,

$$I_2 = \frac{22 \times 10^3}{220} = 100\,\text{A}$$

Example 2.2 A 25 kVA single-phase transformer has the primary and secondary number of turns of 200 and 400, respectively. The transformer is connected to a 220 V, 50 Hz source. Calculate the (i) turns ratio, and (ii) mutual flux in the core.

Solution

(i) The turns ratio is,

$$a = \frac{N_1}{N_2} = \frac{200}{400} = 0.5$$

(ii) The value of the mutual flux can be calculated as,

$$V_1 = E_1 = 4.44 f \phi_m N_1$$
$$\phi_m = \frac{V_1}{4.44 f N_1} = \frac{220}{4.44 \times 50 \times 200} = 4.95\,\text{mWb}$$

Practice problem 2.1
The primary voltage of an iron core single-phase transformer is 220 V. The number of primary and secondary turns of the transformer are 200 and 50, respectively. Calculate the voltage of the secondary coil.

Practice problem 2.2
The number of primary turns of a 30 kVA, 2200/220 V, 50 Hz single-phase transformer is 100. Find the (i) turns ratio, (ii) mutual flux in the core, and (iii) full load primary and secondary currents.

2.7   Rules for Referring Impedance

For developing equivalent circuit of a transformer, it is necessary to refer the parameters from the primary to the secondary or the secondary to the primary. These parameters are resistance, reactance, impedance, current and voltage. The ratio of primary voltage to secondary voltage is [1–4],

$$\frac{V_1}{V_2} = a \tag{2.28}$$

The ratio of primary current to secondary current is,

$$\frac{I_1}{I_2} = \frac{1}{a} \tag{2.29}$$

Dividing Eq. (2.28) by Eq. (2.29) yields,

$$\frac{\frac{V_1}{V_2}}{\frac{I_1}{I_2}} = \frac{a}{\frac{1}{a}} \tag{2.30}$$

$$\frac{\frac{V_1}{I_1}}{\frac{V_2}{I_2}} = a^2 \tag{2.31}$$

$$\frac{Z_1}{Z_2} = a^2 \tag{2.32}$$

Alternative approach: The impedances in the primary and secondary windings are,

$$Z_1 = \frac{V_1}{I_1} \tag{2.33}$$

$$Z_2 = \frac{V_2}{I_2} \tag{2.34}$$

Dividing Eq. (2.33) by Eq. (2.34) yields,

$$\frac{Z_1}{Z_2} = \frac{\frac{V_1}{I_1}}{\frac{V_2}{I_2}} \tag{2.35}$$

$$\frac{Z_1}{Z_2} = \frac{V_1}{V_2} \times \frac{I_2}{I_1} \qquad (2.36)$$

$$\frac{Z_1}{Z_2} = a \times a \qquad (2.37)$$

$$\frac{Z_1}{Z_2} = a^2 \qquad (2.38)$$

From Eq. (2.38), it can be concluded that the impedance ratio is equal to the square of the turns ratio. The important points for transferring parameters are (i) R_1 in the primary becomes $\frac{R_1}{a^2}$ when referred to the secondary, (ii) R_2 in the secondary becomes $a^2 R_2$ when referred to the primary, (iii) X_1 in the primary becomes $\frac{X_1}{a^2}$ when referred to the secondary, and (iv) X_2 in the secondary becomes $a^2 X_2$ when referred to the primary.

Example 2.3 The number of primary and secondary turns of a single-phase transformer are 300 and 30, respectively. The secondary coil is connected with a load impedance of $4\,\Omega$. Calculate the (i) turns ratio, (ii) load impedance referred to the primary, and (iii) primary current if the primary coil voltage is 220 V.

Solution

(i) The value of the turns ratio is,

$$a = \frac{N_1}{N_2} = \frac{300}{30} = 10$$

(ii) The value of the load impedance referred to primary is,

$$Z'_L = a^2 Z_L = 10^2 \times 4 = 400\,\Omega$$

(iii) The value of the primary current is,

$$I_1 = \frac{V_1}{Z'_L} = \frac{220}{400} = 0.55\,A$$

Practice problem 2.3
A load impedance of $8\,\Omega$ is connected to the secondary coil of a 400/200 turns single-phase transformer. Determine the (i) turns ratio, (ii) load impedance referred to primary and (iii) primary current if the primary coil voltage is 120 V.

2.8 Equivalent Circuit of a Transformer

Windings of a transformer are not connected electrically. The windings are magnetically coupled with each other. In this case, it is tedious to do proper analysis. Therefore, for easy computation and visualization, the practical transformer needs to be converted into an equivalent circuit by maintaining same properties of the main transformer. In the equivalent circuit, the related parameters need to be transferred either from the primary to the secondary or vice versa. A two windings ideal transformer is shown in Fig. 2.4.

2.8.1 Exact Equivalent Circuit

Figure 2.5 shows an exact equivalent circuit referred to the primary where all the parameters are transferred from the secondary to the primary and these parameters are,

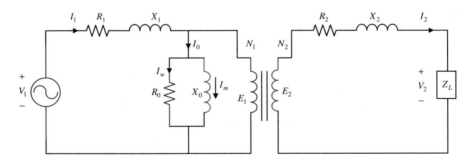

Fig. 2.4 Two windings transformer

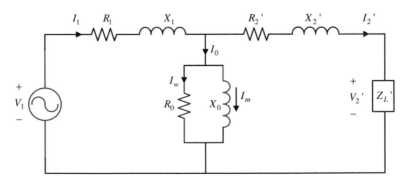

Fig. 2.5 Exact equivalent circuit referred to primary

$$R'_2 = a^2 R_2 \qquad (2.39)$$

$$X'_2 = a^2 X_2 \qquad (2.40)$$

$$Z'_L = a^2 Z_L \qquad (2.41)$$

$$I'_2 = \frac{I_2}{a} \qquad (2.42)$$

$$V'_2 = aV_2 \qquad (2.43)$$

Figure 2.6 shows the exact equivalent circuit referred to the secondary where all the parameters are transferred from the primary to the secondary.

These parameters are,

$$R'_1 = \frac{R_1}{a^2} \qquad (2.44)$$

$$X'_1 = \frac{X_1}{a^2} \qquad (2.45)$$

$$I'_1 = aI_1 \qquad (2.46)$$

$$V'_1 = \frac{V_1}{a} \qquad (2.47)$$

$$I'_w = aI_w \qquad (2.48)$$

$$I'_m = aI_m \qquad (2.49)$$

$$I'_0 = aI_0 \qquad (2.50)$$

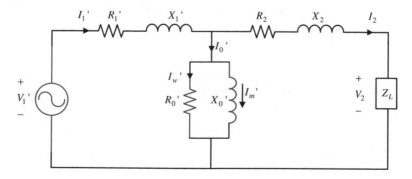

Fig. 2.6 Exact equivalent circuit referred to secondary

2.8.2 Approximate Equivalent Circuit

The no-load current is very small as compared to the rated primary current. Therefore, there is a negligible voltage drop due to R_1 and X_1. In this condition, it can be assumed that the voltage drop across the no-load circuit is the same as the applied voltage without any significant error. The approximate equivalent circuit can be drawn by shifting the no-load circuit across the supply voltage, V_1. Figure 2.7 shows an approximate equivalent circuit referred to the primary. The total resistance, reactance and impedance referred to the primary are,

$$R_{01} = R_1 + R_2' = R_1 + a^2 R_2 \tag{2.51}$$

$$X_{01} = X_1 + X_2' = X_1 + a^2 X_2 \tag{2.52}$$

$$Z_{01} = R_{01} + jX_{01} \tag{2.53}$$

The no-load circuit resistance and reactance are,

$$R_0 = \frac{V_1}{I_w} \tag{2.54}$$

$$X_0 = \frac{V_1}{I_m} \tag{2.55}$$

Figure 2.8 shows an approximate equivalent circuit referred to the secondary. The total resistance, reactance and impedance referred to the secondary is,

$$R_{02} = R_2 + R_1' = R_2 + \frac{R_1}{a^2} \tag{2.56}$$

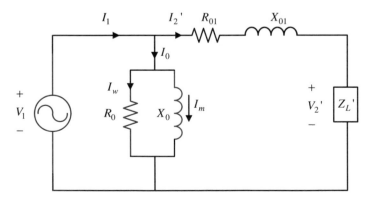

Fig. 2.7 Approximate equivalent circuit referred to primary

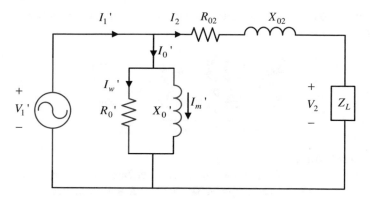

Fig. 2.8 Approximate equivalent circuit referred to secondary

$$X_{02} = X_2 + X_1' = X_2 + \frac{X_1}{a^2} \tag{2.57}$$

$$Z_{02} = R_{02} + jX_{02} \tag{2.58}$$

The no-load circuit resistance and reactance referred to the secondary are,

$$R_0' = \frac{V_1'}{I_w'} \tag{2.59}$$

$$X_0' = \frac{V_1'}{I_m'} \tag{2.60}$$

Example 2.4 A 2.5 kVA, 200 V/40 V single-phase transformer has the primary resistance and reactance of 3 and 12 Ω, respectively. On the secondary side, these values are 0.3 and 0.1 Ω, respectively. Find the equivalent impedance referred to the primary and the secondary.

Solution

The value of the turns ratio is,

$$a = \frac{V_1}{V_2} = \frac{200}{40} = 5$$

The total resistance, reactance and impedance referred to the primary can be determined as,

$$R_{01} = R_1 + a^2 R_2 = 3 + 25 \times 0.3 = 10.5\,\Omega$$
$$X_{01} = X_1 + a^2 X_2 = 12 + 25 \times 0.1 = 14.5\,\Omega$$
$$Z_{01} = \sqrt{10.5^2 + 14.5^2} = 17.9\,\Omega$$

The total resistance, reactance and impedance referred to the secondary are calculated as,

$$R_{02} = R_2 + \frac{R_1}{a^2} = 0.3 + \frac{3}{25} = 0.42\,\Omega$$
$$X_{02} = X_2 + \frac{X_1}{a^2} = 0.1 + \frac{12}{25} = 0.58\,\Omega$$
$$Z_{01} = \sqrt{0.42^2 + 0.58^2} = 0.72\,\Omega$$

Practice problem 2.4
A 1.5 kVA, 220/110 V single-phase transformer has the primary resistance and reactance of 6 and 18 Ω, respectively. The resistance and reactance at the secondary side are 0.6 and 0.5 Ω, respectively. Calculate the equivalent impedance referred to the primary and the secondary.

2.9 Polarity of a Transformer

The relative directions of induced voltages between the high voltage and low voltage terminals is known as the polarity of a transformer. The polarity of a transformer is very important to construct three-phase transformer bank, parallel connection of transformer, connection of current transformer (CT) and potential transformer (PT) power with metering device. Two polarities namely additive and subtractive are used in the transformer.

A polarity of a transformer is said to be an additive if the measured voltage between the high voltage and the low voltage terminals is greater than the supply voltage at the high voltage terminals. The additive polarity of a transformer is marked by the orientation of dots as shown in Fig. 2.9. Whereas, a polarity is said to be a subtractive if the measured voltage between the high voltage and the low voltage terminals is lower than the supply voltage at the high voltage terminals. The

Fig. 2.9 Additive polarity

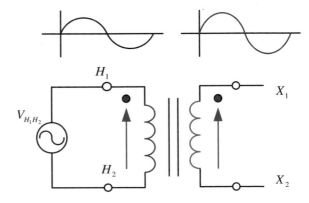

Fig. 2.10 Subtractive
polarity

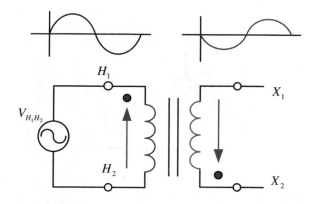

subtractive polarity of a transformer is marked by the orientation of dots as shown
in Fig. 2.10. Consider a 220/110 V single-phase transformer with the high voltage
and the low voltage terminals for testing polarities. The high voltage terminal H_1 is
connected to the low voltage terminal X_1 by a cable. The voltmeter is connected
between H_2 and X_2. In this case, the turns ratio of the transformer is,

$$a = \frac{V_1}{V_2} = \frac{220}{110} = 2$$

For safety issue, a lower voltage needs to be applied to the primary side i.e., high
voltage terminals. Suppose, a voltage of 110 V is applied to the primary side. In this
case, a voltage of 55 V (110/2) will appear at the secondary terminals. If the meter
read out the voltage of 165 V (110 + 55) then the transformer is said to be in
additive polarity. This connection is shown in Fig. 2.11.

Whereas, if the voltmeter reads the voltage of 55 V (110 − 55) then the transformer is said to be in subtractive polarity as shown in Fig. 2.12.

Fig. 2.11 Testing of additive
polarity

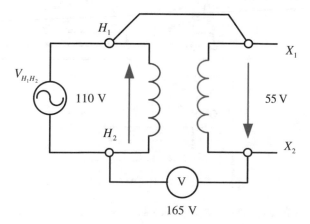

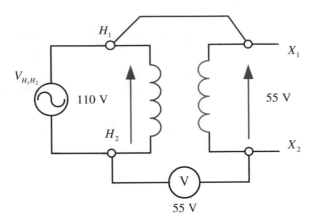

Fig. 2.12 Testing of subtractive polarity

2.10 Three-Phase Transformer

A three-phase power transformer is used at the power generating station to step-up the voltage from 11 to 120 kV. Whereas in the power distribution substation, the three-phase voltage is again stepped down to 11 kV voltage through a three-phase distribution transformer.

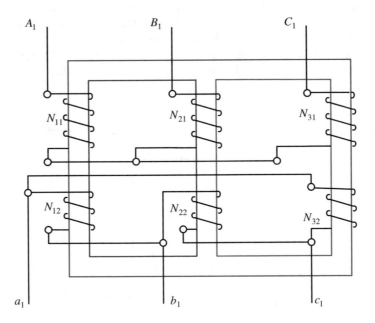

Fig. 2.13 Three windings on a common core and wye-delta connection

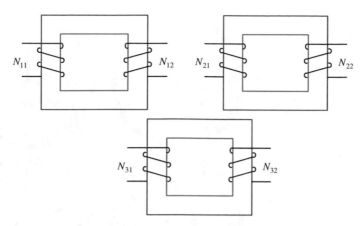

Fig. 2.14 Three single-phase transformers

Therefore, a three-phase transformer can be made either by three windings wound on a common core or by three single-phase transformer connected together in a three-phase bank.

The first approach is a cheaper one that results in a transformer with smaller size and less weight. The main disadvantage of the first approach is that if one phase becomes defective, then the whole transformer needs to be replaced. Whereas in the second approach, if one of the transformers becomes defective then the system can be given power by an open delta at a reduced capacity. In this case, the defective transformer is normally replaced by a new one. A three-phase transformer with wye-delta connection is shown in Fig. 2.13 and the three single-phase transformers are shown in Fig. 2.14.

2.11   Transformer Vector Group

The most common configurations of a three-phase transformer are the delta and star configurations, which are used in the power utility company. The primary and secondary windings of a three-phase transformer are connected either in the same (delta-delta or star-star), or different (delta-star or star-delta) configuration-pair. The secondary voltage waveforms of a three-phase transformer are in phase with the primary waveforms when the primary and secondary windings are connected in the same configuration. This condition is known as 'no phase shift' condition. If the primary and secondary windings are connected in different configuration pair then the secondary voltage waveforms will differ from the corresponding primary

Fig. 2.15 Representation of vector groups with clock

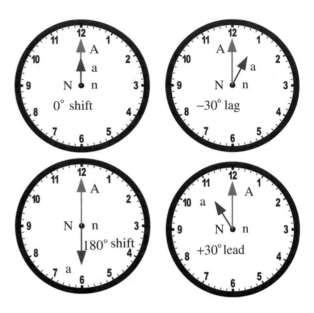

voltage waveforms by 30 electrical degrees. This condition is called a '30° phase shift' condition. The windings and their position to each other are usually marked by vector group. The vector group is used to identify the phase shift between the primary and secondary windings. In the vector group, the secondary voltage may have the phase shift of 30° lagging or leading, 0° i.e., no phase shift or 180° reversal with respect to the primary voltage. The transformer vector group is labeled by capital and small letters plus numbers from 1 to 12 in a typical clock-like diagram. The capital letter indicates primary winding and small letter represents secondary winding. In the clock diagram, the minute hand represents the primary line to neutral line voltage, and its place is always in the 12. The hour hand represents the secondary line to neutral voltage and its position in the clock changes based on the phase shift as shown in Fig. 2.15.

There are four vector groups used in the three-phase transformer connection. These vector groups are (i) Group I: 0 o'clock, zero phase displacement (Yy0, Dd0, Dz0), (ii) Group II: 6 o'clock, 180° phase displacement (Yy6, Dd6, Dz6), (iii) Group III: 1 o'clock, −30° lag phase displacement (Dy1, Yd1, Yz1), and (iv) Group IV: 11 o'clock, 30° lead phase displacement (Dy11, Yd11, Yz11). Here, Y represents wye connection, D represents delta connection and z represents the zigzag connection. The connection diagrams for different combinations are shown in Figs. 2.16, 2.17, 2.18, 2.19, 2.20, 2.21, 2.22, 2.23, 2.24, 2.25, 2.26 and 2.27.

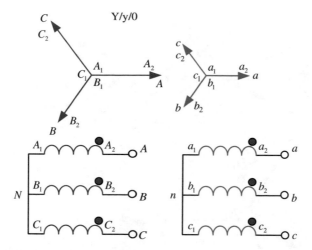

Fig. 2.16 Connection of Y-y-0

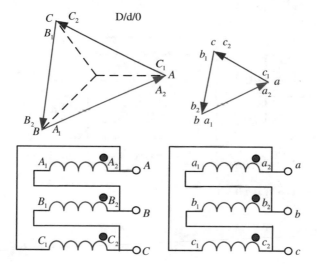

Fig. 2.17 Connection of D-d-0

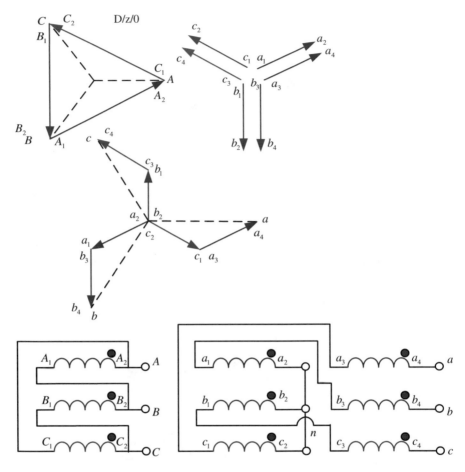

Fig. 2.18   Connection of D-z-0

Fig. 2.19   Connection of
Y-d-1

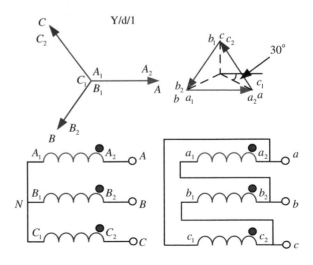

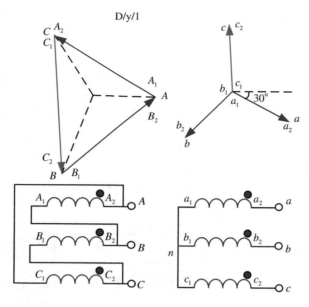

Fig. 2.20   Connection of D-y-1

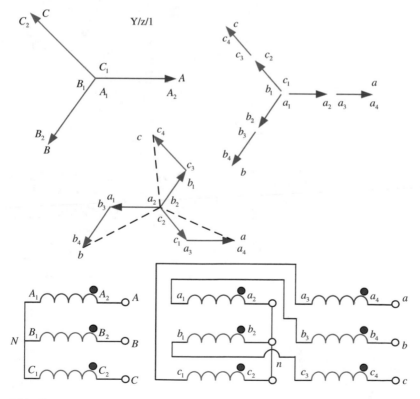

Fig. 2.21   Connection of Y-z-1

Fig. 2.22 Connection of D-y-11

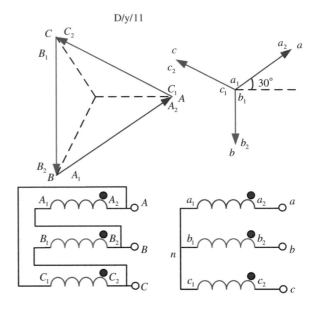

Fig. 2.23 Connection of Y-d-11

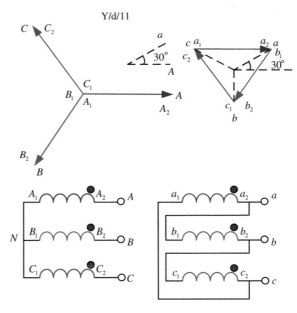

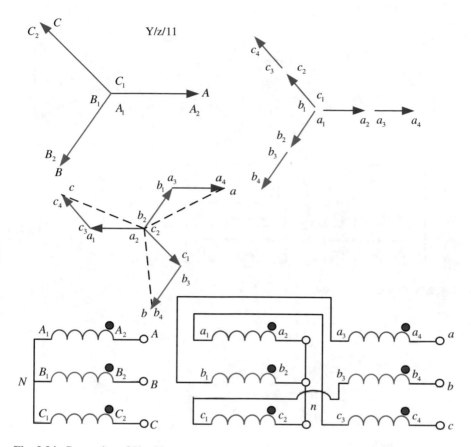

Fig. 2.24 Connection of Y-z-11

Fig. 2.25 Connection of Y-y-6

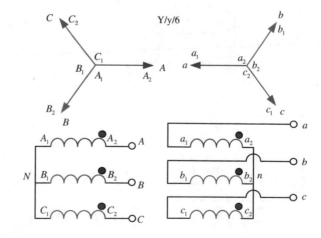

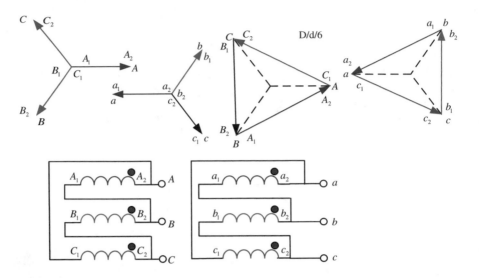

Fig. 2.26   Connection of D-d-6

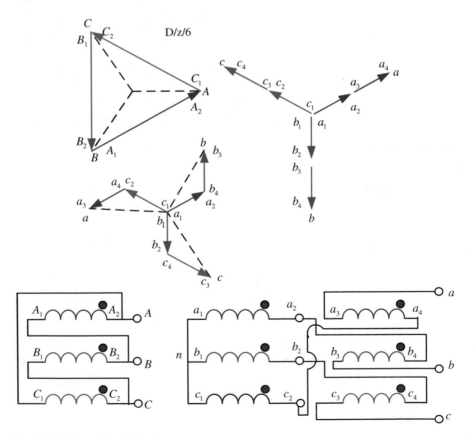

Fig. 2.27   Connection of D-y-6

2.12   Voltage Regulation of a Transformer

Different types of loads like domestic, commercial and industrial are usually connected with the secondary winding of a transformer. All these loads are operated
with a constant magnitude of voltage. The secondary voltage of a transformer under
operation changes due to voltage drop across the internal impedance and the load.
The voltage regulation of a transformer is used to identify the characteristic of the
secondary side voltage changes under different loading conditions. The voltage
regulation of a transformer is defined as the difference between the no-load terminal
voltage (V_{2NL}) to full load terminal voltage (V_{2FL}) and is expressed as a percentage
of full load terminal voltage. It is therefore can be expressed as,

$$\text{Voltage regulation} = \frac{V_{2NL} - V_{2FL}}{V_{2FL}} \times 100\,\% = \frac{E_2 - V_2}{V_2} \times 100\,\% \qquad (2.61)$$

Figure 2.28 shows an approximate equivalent circuit referred to the secondary
without no-load circuit to find the voltage regulation for different power factors.
Phasor diagrams for different power factors are shown in Fig. 2.29 where the
secondary voltage V_2 is considered as the reference phasor.

The phasor diagram with a unity power factor is shown in Fig. 2.29a and the
phasor form of the secondary induced voltage for a unity power factor can be
written as,

$$E_2 = V_2 + I_2(R_{02} + jX_{02}) \qquad (2.62)$$

Figure 2.29b shows the phasor diagram for a lagging power factor, and from this
diagram, the following expressions can be written,

$$AC = V_2 \cos \phi_2 \qquad (2.63)$$

$$BC = DE = V_2 \sin \phi_2 \qquad (2.64)$$

$$CD = BE = I_2 R_{02} \qquad (2.65)$$

$$EF = I_2 X_{02} \qquad (2.66)$$

Fig. 2.28 Approximate
equivalent circuit referred to
secondary

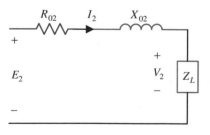

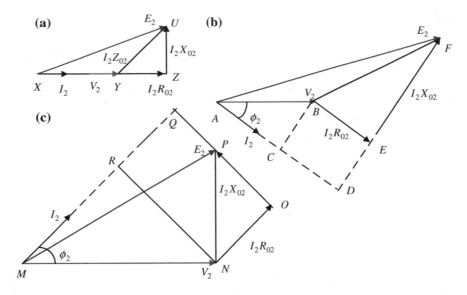

Fig. 2.29 Phasor diagram for different power factors

From the right angle triangle-*ADF*, the expression of E_2 can be derived as,

$$AF^2 = AD^2 + DF^2 \tag{2.67}$$

$$AF^2 = (AC + CD)^2 + (DE + EF)^2 \tag{2.68}$$

Substituting equations from (2.63) to (2.66) into Eq. (2.68) yields,

$$E_2 = \sqrt{(V_2 \cos \phi_2 + I_2 R_{02})^2 + (V_2 \sin \phi_2 + I_2 X_{02})^2} \tag{2.69}$$

In phasor form, Eq. (2.69) can be written as,

$$E_2 = (V_2 \cos \phi_2 + I_2 R_{02}) + j(V_2 \sin \phi_2 + I_2 X_{02}) \tag{2.70}$$

Figure 2.29c shows the phasor diagram for a leading power factor and from this diagram, the following expressions can be written,

$$MR = V_2 \cos \phi_2 \tag{2.71}$$

$$RN = OQ = V_2 \sin \phi_2 \tag{2.72}$$

$$RQ = NO = I_2 R_{02} \tag{2.73}$$

$$PO = I_2 X_{02} \tag{2.74}$$

The expression of E_2 from the right angled triangle MQP can be derived as,

$$MP^2 = MQ^2 + QP^2 \tag{2.75}$$

$$MP^2 = (MR + RQ)^2 + (OQ - PO)^2 \tag{2.76}$$

Substituting equations from (2.71) to (2.74) into Eq. (2.76) yields,

$$E_2 = \sqrt{(V_2 \cos \phi_2 + I_2 R_{02})^2 + (V_2 \sin \phi_2 - I_2 X_{02})^2} \tag{2.77}$$

In phasor form, Eq. (2.77) can be written as,

$$E_2 = (V_2 \cos \phi_2 + I_2 R_{02}) + j(V_2 \sin \phi_2 - I_2 X_{02}) \tag{2.78}$$

Example 2.5 The primary coil resistance and reactance of a 200/400 V single-phase transformer are 0.3 and 0.6 Ω, respectively. The secondary coil resistance and reactance are 0.8 and 1.6 Ω, respectively. Calculate the voltage regulation if the secondary current of the transformer is 10 A at a 0.8 pf lagging.

Solution

The value of the turns ratio is,

$$a = \frac{V_1}{V_2} = \frac{200}{400} = 0.5$$

The value of the total resistance referred to the secondary is,

$$R_{02} = R_2 + \frac{R_1}{a^2} = 0.8 + \frac{0.3}{0.25} = 2\,\Omega$$

The total reactance referred to the primary is,

$$X_{02} = X_2 + \frac{X_1}{a^2} = 1.6 + \frac{0.6}{0.25} = 4\,\Omega$$

The no-load voltage is,

$$E_2 = (V_2 \cos \phi_2 + I_2 R_{02}) + j(V_2 \sin \phi_2 + I_2 X_{02})$$
$$E_2 = (400 \times 0.8 + 10 \times 2) + j(400 \times 0.6 + 40)$$
$$E_2 = 440.5 \underline{|39.5^\circ}\ \text{V}$$

The voltage regulation can be determined as,

$$\text{Voltage regulation} = \frac{E_2 - V_2}{V_2} \times 100\%$$

$$\text{Voltage regulation} = \frac{440.5 - 400}{400} \times 100\% = 10\%$$

Practice problem 2.5

A 110/220 V single-phase transformer has the resistance of $0.2\,\Omega$ and a reactance of $0.8\,\Omega$ in the primary winding. The resistance and reactance in the secondary winding are 0.9 and $1.8\,\Omega$, respectively. Calculate the voltage regulation, when the secondary current is 6 A at a 0.85 power factor leading.

2.13 Efficiency of a Transformer

The efficiency is an important parameter to identify the characteristics of any machine. The efficiency η, of any machine can be defined as the ratio of its output to the input. Mathematically, it can be expressed as,

$$\eta = \frac{\text{output}}{\text{input}} = \frac{\text{input} - \text{losses}}{\text{input}} = 1 - \frac{\text{losses}}{\text{input}} \tag{2.79}$$

Let us consider the following:

P_{out} is the output power, in W,
P_{in} is the input power, in W,
P_{losses} is the total losses, in W.

Equation (2.79) can be modified as,

$$\eta = \frac{P_{out}}{P_{in}} = 1 - \frac{P_{losses}}{P_{in}} \tag{2.80}$$

It is worth nothing that the efficiency of a transformer is generally higher than other electrical machines because the transformer has no moving parts.

2.14 Iron and Copper Losses

The iron loss of a transformer is often called as core loss, which is a result of an alternating flux in the core of the transformer. The iron loss consists of the eddy current loss and the hysteresis loss. In the transformer, most of the flux transferred from the primary coil to the secondary coil through low reluctance iron path. A few portion of that alternating flux links with the iron parts of the transformer. As a result, an emf is induced in the transformer core. A current will flow in that parts of the transformer. This current does not contribute in output of the transformer but dissipated as heat. This current is known as eddy current and the power loss due to this current is known as eddy current loss. The eddy current loss (P_e) is directly proportional to the square of the frequency (f) times the maximum magnetic flux density (B_m) and the eddy current loss can be expressed as,

$$P_e = k_e f^2 B_m^2 \qquad (2.81)$$

where k_e is the proportionality constant.

Steel is a very good ferromagnetic material which is used for the core of a transformer. This ferromagnetic material contains number of domains in the structure and magnetized easily. These domains are like small magnets located randomly in the structure. When an mmf is applied to the core then those domains change their position. After removing mmf, most of the domains come back to their original position and remaining will be as it is. As a result, the substance is slightly permanently magnetized. An additional mmf is required to change the position of the remaining domains. Therefore, hysteresis loss is defined as the additional energy is required to realign the domains in the ferromagnetic material. The hysteresis loss (P_h) is directly proportional to the frequency (f) and 2.6th power of the maximum magnetic flux density (B_m) and the expression of hysteresis loss is,

$$P_h = k_h f B_m^{2.6} \qquad (2.82)$$

where k_e is the proportionality constant.

Now that the magnetic flux density is usually constant, Eqs. (2.81) and (2.82) can be modified as,

$$P_e \propto f^2 \qquad (2.83)$$

$$P_h \propto f \qquad (2.84)$$

Practically, hysteresis loss depends on the voltage and the eddy current loss depends on the current. Therefore, total losses of the transformer depend on the voltage and the current not on the power factor. That is why the transformer rating is always represented in kVA instead of kW.

In the transformer, copper losses occur due to the primary and the secondary resistances. The full load copper losses can be determined as,

$$P_{culoss} = I_1^2 R_1 + I_2^2 R_2 \qquad (2.85)$$

$$P_{culoss} = I_1^2 R_{01} = I_2^2 R_{02} \qquad (2.86)$$

2.15 Condition for Maximum Efficiency

The expression of the output power of a transformer is,

$$P_{out} = V_2 I_2 \cos \phi_2 \tag{2.87}$$

The expression of the copper loss is,

$$P_{cu} = I_2^2 R_{02} \tag{2.88}$$

The expression of an iron loss is,

$$P_{iron} = P_{eddy} + P_{hys} \tag{2.89}$$

According to Eq. (2.79), the efficiency can be expressed as,

$$\eta = \frac{\text{output}}{\text{input}} = \frac{\text{output}}{\text{output} + \text{losses}} \tag{2.90}$$

Substituting Eqs. (2.87), (2.88) and (2.89) into Eq. (2.90) yields,

$$\eta = \frac{V_2 I_2 \cos \phi_2}{V_2 I_2 \cos \phi_2 + P_{iron} + P_{cu}} \tag{2.91}$$

$$\eta = \frac{V_2 \cos \phi_2}{V_2 \cos \phi_2 + \frac{P_{iron}}{I_2} + \frac{I_2^2 R_{02}}{I_2}} \tag{2.92}$$

The terminal voltage at the secondary side is considered to be constant in case of a normal transformer. For a given power factor, the secondary current is varied with the variation of load. Therefore, the transformer efficiency will be maximum, if the denominator of Eq. (2.92) is minimum. The denominator of Eq. (2.92) will be minimum, for the following condition.

$$\frac{d}{dI_2} \left(V_2 \cos \phi_2 + \frac{P_{iron}}{I_2} + \frac{I_2^2 R_{02}}{I_2} \right) = 0 \tag{2.93}$$

$$-\frac{P_{iron}}{I_2^2} + R_{02} = 0 \tag{2.94}$$

$$P_{iron} = I_2^2 R_{02} \tag{2.95}$$

From Eq. (2.95), it is concluded that the efficiency of a transformer will be maximum, when the iron loss is equal to the copper loss. From Eq. (2.95), the expression of secondary current can be written as,

$$I_2 = \sqrt{\frac{P_{\text{iron}}}{R_{02}}} \tag{2.96}$$

For maximum efficiency, the load current can be expressed as,

$$I_{2\eta} = \sqrt{\frac{P_{\text{iron}}}{R_{02}}} \tag{2.97}$$

Equation (2.97) can be re-arranged as,

$$I_{2\eta} = I_2 \sqrt{\frac{P_{\text{iron}}}{I_2^2 R_{02}}} \tag{2.98}$$

Multiplying both sides of Eq. (2.98) by the secondary rated voltage V_2 yields,

$$V_2 I_{2\eta} = V_2 I_2 \sqrt{\frac{P_{\text{iron}}}{I_2^2 R_{02}}} \tag{2.99}$$

$$VI_{\text{max efficiency}} = VI_{\text{rated}} \times \sqrt{\frac{\text{iron loss}}{\text{full load copper loss}}} \tag{2.100}$$

Example 2.6 A 30 kVA transformer has the iron loss and full load copper loss of 350 and 650 W, respectively. Determine the (i) full load efficiency, (ii) output kVA corresponding to maximum efficiency, and (iii) maximum efficiency. Consider that the power factor of the load is 0.6 lagging.

Solution

(i) The total value of full load loss is,

$$P_{tloss} = 350 + 650 = 1000\,\text{W} = 1\,\text{kW}$$

The output power at full load is,

$$P_{out} = 30 \times 0.6 = 18\,\text{kW}$$

The input power at full load is,

$$P_{in} = 18 + 1 = 19\,\text{kW}$$

The efficiency at full load is,

$$\eta = \frac{18}{19} \times 100 = 94.74\%$$

(ii)   The output kVA corresponding to maximum efficiency is,

$$= kVA_{rated} \times \sqrt{\frac{\text{Iron loss}}{\text{Cu loss at full load}}}$$

$$= 30 \times \sqrt{\frac{350}{650}} = 22\,kVA$$

The output power is,

$$P_{o1} = 22 \times 0.6 = 13.2\,kW$$

(iii) For maximum efficiency, iron loss is equal to copper loss.
The total value of the loss is,

$$P_{tloss-1} = 2 \times 350 = 700\,W$$

The value of the input power is,

$$P_{in1} = 13.2 + 0.7 = 13.9\,kW$$

The efficiency is,

$$\eta = \frac{13.2}{13.9} \times 100 = 94.96\%$$

Practice problem 2.6
A 200 kVA transformer is having an iron loss of 1.5 kW and a copper loss of 5 kW
at full load condition. Find the kVA rating at which the efficiency is maximum.
Also, find the efficiency at a unity power factor.

2.16   Transformer Tests

The equivalent circuit parameters are very important to characterize the perfor-
mance of transformer. The parameters of a transformer equivalent circuit can be
determined by the open circuit and the short circuit tests.

2.16.1   Open Circuit Test

The main objectives of the open circuit test are to determine the no-load current and iron loss. The components of the no-load current are used to determine the no-load circuit resistance and reactance.

In an open circuit test, the secondary side is considered to be open circuit, and the primary coil is connected to the source as shown in Fig. 2.30a, where all measuring instruments are connected in the primary side. A specific alternating voltage is applied to the primary winding. Then the wattmeter will measure the iron loss and small amount of copper loss. The ammeter and voltmeter will measure the no-load current and the voltage, respectively. Since, the no-load current is very small, the copper losses can be neglected. Then the wattmeter reading can be expressed as,

$$P_0 = V_1 I_0 \cos \phi_0 \tag{2.101}$$

From Eq. (2.101), the no-load power factor can be determined as,

$$\cos \phi_0 = \frac{P_0}{V_1 I_0} \tag{2.102}$$

The working and magnetizing components of the current can be determined as,

$$I_w = I_0 \cos \phi_0 \tag{2.103}$$

$$I_m = I_0 \sin \phi_0 \tag{2.104}$$

Then the no-load circuit resistance and reactance can be determined as,

$$R_0 = \frac{V_1}{I_w} \tag{2.105}$$

$$X_0 = \frac{V_1}{I_m} \tag{2.106}$$

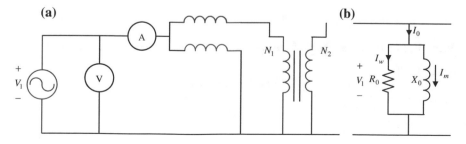

Fig. 2.30 Connection diagrams for open circuit test and no-load circuit

Example 2.7 A 200/400 V, 50 Hz single-phase transformer has the no-load test data of 200 V, 0.6 A, 80 W. Calculate the no-load circuit resistance and reactance.

Solution

The power factor can be determined as,

$$\cos \phi_0 = \frac{P_0}{V_1 I_0} = \frac{80}{200 \times 0.6} = 0.67$$
$$\sin \phi_0 = 0.74$$

The values of the working and magnetizing components of the no-load current are,

$$I_w = I_0 \cos \phi_0 = 0.6 \times 0.67 = 0.4 \, \text{A}$$
$$I_m = I_0 \sin \phi_0 = 0.6 \times 0.74 = 0.44 \, \text{A}$$

The no-load circuit parameters can be determined as,

$$R_0 = \frac{V_1}{I_w} = \frac{200}{0.4} = 500 \, \Omega$$
$$X_0 = \frac{V_1}{I_m} = \frac{200}{0.44} = 454.5 \, \Omega$$

2.16.2 Short Circuit Test

The main objectives of the short circuit test are to determine the equivalent resistance, reactance, impedance and full load copper loss. In this test, the supply voltage and the measuring instruments (e. g,. wattmeter, ammeter) are connected to the primary side and the secondary winding is shorted out by a wire as shown in Fig. 2.31, or connected with an ammeter. The primary voltage is adjusted until the current in the short-circuited winding is equal to the rated primary current. Under this condition, the wattmeter will measure the full load copper loss and it can be written as,

$$P_{sc} = I_{sc} V_{sc} \cos \phi_{sc} \tag{2.107}$$

From Eq. (2.107), the short circuit power factor can be calculated as,

$$\cos \phi_{sc} = \frac{P_{sc}}{I_{sc} V_{sc}} \tag{2.108}$$

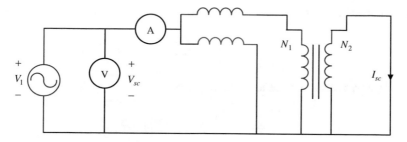

Fig. 2.31 Connection diagram for short circuit test

The equivalent impedance can be calculated as,

$$Z_{01} = Z_{eq} = \frac{V_{sc}}{I_{sc}} \qquad (2.109)$$

In addition, the equivalent resistance and reactance can be calculated as,

$$R_{01} = Z_{01} \cos \phi_{sc} \qquad (2.110)$$

$$X_{01} = Z_{01} \sin \phi_{sc} \qquad (2.111)$$

Example 2.8 A 25 kVA, 2200/220 V, 50 Hz single-phase transformer's low voltage side is short-circuited and the test data recorded from the high voltage side are $P = 150$ W, $I_1 = 5$ A and $V_1 = 40$ V. Determine the (i) equivalent resistance, reactance and impedance referred to primary, (ii) equivalent resistance, reactance and impedance referred to secondary, and (iii) voltage regulation at unity power factor.

Solution

(i) The parameters referred to primary are,

$$Z_{01} = \frac{V_1}{I_1} = \frac{40}{5} = 8\,\Omega$$

$$R_{01} = \frac{P}{I_1^2} = \frac{150}{25} = 6\,\Omega$$

$$\phi = \cos^{-1}\left(\frac{6}{8}\right) = 41.4°$$

$$X_{01} = Z_{01} \sin \phi = 8 \sin 41.4° = 5.2\,\Omega$$

(ii)   The turns ratio is,

$$a = \frac{V_1}{V_2} = \frac{2200}{220} = 10$$

The parameters referred to secondary are,

$$Z_{02} = \frac{Z_{01}}{a^2} = \frac{8}{100} = 0.08\,\Omega$$

$$R_{02} = \frac{R_{01}}{a^2} = \frac{6}{100} = 0.06\,\Omega$$

$$X_{02} = \frac{X_{01}}{a^2} = \frac{5.2}{100} = 0.052\,\Omega$$

(iii)   The secondary side current is,

$$I_2 = \frac{25{,}000}{220} = 113.6\,\text{A}$$

The secondary induced voltage is,

$$E_2 = V_2 + I_2 Z_{02} = 220 + 113.6 \times 0.08 = 229\,\text{V}$$

The voltage regulation is,

$$VR = \frac{E_2 - V_2}{V_2} = \frac{229 - 220}{220} = 4\,\%$$

Practice problem 2.7
The no-load test data of a 220/2200 V, 50 Hz single-phase transformer are 220 V, 0.4 A, 75 W. Calculate the no-load circuit resistance and reactance.

Practice problem 2.8
A 30 kVA, 2300/220 V, 50 Hz single-phase transformer's low voltage side is short-circuited by a thick wire. The test data recorded from the high voltage side are $P = 400\,\text{W}$, $I_1 = 8.5\,\text{A}$ and $V_1 = 65\,\text{V}$. Find the (i) equivalent resistance, reactance and impedance referred to primary and (ii) equivalent resistance, reactance and impedance referred to secondary, and (iii) voltage regulation at a 0.6 lagging power factor.

2.17   Autotransformer

A small rating transformer with a variable voltage output is usually used in the educational laboratory as well as in the testing laboratory. This type of small rating transformer with a variable output is known as autotransformer. An autotransformer has one continuous winding that is common to both the primary and the secondary. Therefore, in an autotransformer, the primary and secondary windings are connected electrically. The advantages of an autotransformer over a two-winding transformer include lower initial investment, lower leakage reactance, lower losses compared to conventional transformer and lower excitation current.

An autotransformer with primary and secondary windings is shown in Fig. 2.32. In this connection, the suffix c indicates the common winding and the suffix s indicates the series winding. From Fig. 2.32, the following equations can be written as,

$$\frac{N_s}{N_c} = \frac{I_c}{I_s} = a \qquad (2.112)$$

$$\frac{V_s}{V_c} = \frac{N_s}{N_c} = a \qquad (2.113)$$

The total voltage in the primary side is,

$$V_1 = V_s + V_c = V_c\left(1 + \frac{V_s}{V_c}\right) \qquad (2.114)$$

Substituting Eq. (2.113) into Eq. (2.114) yields,

$$\frac{V_1}{V_c} = (1 + a) \qquad (2.115)$$

$$\frac{V_L}{V_1} = \frac{1}{1 + a} \qquad (2.116)$$

Fig. 2.32 Connection diagram for an autotransformer

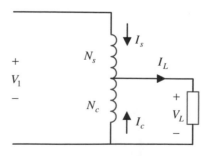

Where the voltage at the load is equal to the voltage at the common terminals i.e., $V_L = V_c$.

The expression of the load current can be written as,

$$I_L = I_s + I_c = aI_s + I_s \tag{2.117}$$

$$I_L = (1+a)I_s \tag{2.118}$$

$$\frac{I_L}{I_s} = (1+a) \tag{2.119}$$

Example 2.9 Figure 2.33 shows a single-phase 120 kVA, 2200/220 V, 50 Hz transformer which is connected as an autotransformer. The voltages of the upper and lower parts of the coil are 220 and 2200 V, respectively. Calculate the kVA rating of the autotransformer.

Solution

The current ratings of the respective windings are,

$$I_{pq} = \frac{120,000}{220} = 545.5\,\text{A}$$
$$I_{qr} = \frac{120,000}{2200} = 54.5\,\text{A}$$

The current in the primary side is,

$$I_1 = 545.5 + 54.5 = 600\,\text{A}$$

The voltage across the secondary side is,

$$V_2 = 2200 + 220 = 2420\,\text{V}$$

Fig. 2.33 An autotransformer with a specific voltage

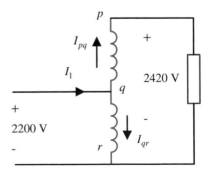

Fig. 2.34 An
autotransformer with a
specific voltage

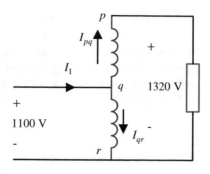

Therefore, the kVA ratings of an autotransformer,

$$kVA_1 = \frac{600 \times 2200}{1000} = 1320\,kVA$$

$$kVA_2 = \frac{545.5 \times 2420}{1000} = 1320\,kVA$$

Practice problem 2.9
A single-phase 100 kVA, 1100/220 V, 50 Hz transformer is connected as an autotransformer as shown in Fig. 2.34. The voltages of the upper and lower parts of the coil are 220 and 1100 V, respectively. Determine the kVA rating of the autotransformer.

2.18   Parallel Operation of a Single-Phase Transformer

Nowadays, the demand of load is increasing with the increase in population and industrial sector. Sometimes, it is difficult to meet the excess demand of power by the existing single unit transformer. Therefore, an additional transformer is required to connect in parallel with the existing one. The following points should be considered for making parallel connections of transformers:

- Terminal voltage of both transformers must be the same.
- Polarity must be same for both transformers.
- For both transformers, the percentage impedances should be equal in magnitude.
- Ratio of R to X must be the same for both transformers.
- Phase sequences and phase shifts must be the same (for three-phase transformer).

2.19 Three-Phase Transformer Connections

The primary and secondary windings of the transformer may be connected in either by wye (Y) or delta (Δ). Three-phase transformer connections are classified into four possible types; namely, Y-Y(wye-wye), Y-Δ (wye-delta), Δ-Y(delta-wye) and Δ-Δ (delta-delta).

2.19.1 Wye-Wye Connection

Figure 2.35 shows the Y-Y connection diagram. This type of connection of a three-phase transformer is rarely used for large amount of power transmission.

Neutral point is necessary for both primary and secondary sides in some cases. In balanced loads, this type of connection works satisfactorily and provides neutral to each side for grounding. At the primary side, the phase voltage can be written as,

$$V_{P1} = \frac{V_{L1}}{\sqrt{3}} \tag{2.120}$$

The secondary phase voltage can be written as,

$$V_{P2} = \frac{V_{L2}}{\sqrt{3}} \tag{2.121}$$

The ratio of the primary line voltage to the secondary line voltage of this connection is,

$$a = \frac{V_{L1}}{V_{L2}} \tag{2.122}$$

$$a = \frac{\sqrt{3}V_{P1}}{\sqrt{3}V_{P2}} = \frac{V_{P1}}{V_{P2}} \tag{2.123}$$

Fig. 2.35 Y-Y connection diagram

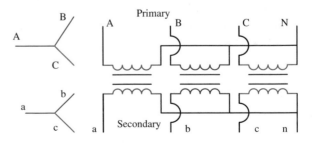

2.19.2 Wye-Delta Connection

The wye-delta connection is mainly used at the substation where the voltage is stepped down. In this connection, the current in the secondary coil is 57.7 % of the load current. At the primary side of this connection, a proper copper wire is used to ground the neutral point. Figure 2.36 shows the connection diagram of the wye-delta transformer. In this connection, the expression of the primary line voltage is,

$$V_{L1} = \sqrt{3}V_{P1} \tag{2.124}$$

At the secondary side, the line voltage is,

$$V_{L2} = V_{P2} \tag{2.125}$$

The ratio of primary phase voltage to secondary phase voltage is,

$$a = \frac{V_{P1}}{V_{P2}} \tag{2.126}$$

The ratio of primary line voltage to secondary line voltage is,

$$\frac{V_{L1}}{V_{L2}} = \frac{\sqrt{3}V_{P1}}{V_{P2}} = \sqrt{3}a \tag{2.127}$$

The primary phase current is,

$$I_{P1} = I_{L1} \tag{2.128}$$

In this case, the turns ratio is,

$$a = \frac{I_{P2}}{I_{P1}} \tag{2.129}$$

Fig. 2.36 Wye-delta connection diagram

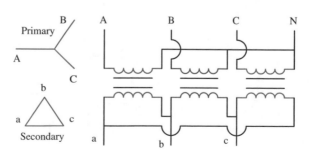

The expression of the secondary phase current is,

$$I_{P2} = aI_{P1} \tag{2.130}$$

The secondary line current is,

$$I_{L2} = \sqrt{3}I_{P2} = \sqrt{3}aI_{P1} \tag{2.131}$$

2.19.3 Delta-Wye Connection

The delta-wye connection is generally used at the power generating station to step the voltage. Figure 2.37 shows the connection diagram of a delta-wye transformer. In this connection, the expression of the primary line voltage is,

$$V_{L1} = V_{P1} \tag{2.132}$$

The line voltage at the secondary side is,

$$V_{L2} = \sqrt{3}V_{P2} \tag{2.133}$$

The ratio of primary line voltage to secondary line voltage is,

$$\frac{V_{L1}}{V_{L2}} = \frac{V_{P1}}{\sqrt{3}V_{P2}} = \frac{a}{\sqrt{3}} \tag{2.134}$$

The phase current at the primary side is,

$$I_{P1} = \frac{1}{\sqrt{3}}I_{L1} \tag{2.135}$$

Fig. 2.37 Delta-wye connection diagram

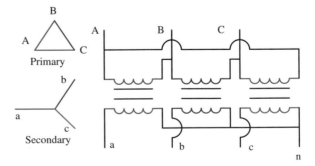

In this connection, the turns ratio is,

$$a = \frac{I_{P2}}{I_{P1}} \tag{2.136}$$

In this case, the secondary phase current is,

$$I_{P2} = aI_{P1} \tag{2.137}$$

The secondary line current is,

$$I_{L2} = I_{P2} = aI_{P1} \tag{2.138}$$

2.19.4   Delta-Delta Connection

The delta-delta connection is generally used for both high voltage and low voltage rating transformers where insulation is not an important issue. The connection diagram for delta-delta transformer is shown in Fig. 2.38. In this case, the primary line voltage is,

$$V_{L1} = V_{P1} \tag{2.139}$$

The secondary line voltage is,

$$V_{L2} = V_{P2} \tag{2.140}$$

The ratio of line-to-line voltage of this connection is,

$$\frac{V_{L1}}{V_{L2}} = \frac{V_{P1}}{V_{P2}} = a \tag{2.141}$$

Fig. 2.38 Delta-delta connection diagram

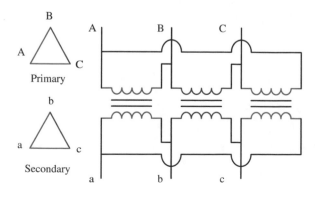

The secondary line current is,

$$I_{L2} = \sqrt{3}\, I_{P2} \qquad\qquad (2.142)$$

The output capacity in delta-delta connection can be expressed as,

$$P_{0\Delta-\Delta} = \sqrt{3}V_{L2}I_{L2} \qquad\qquad (2.143)$$

Substituting Eq. (2.142) into Eq. (2.43) yields,

$$P_{0\Delta-\Delta} = \sqrt{3}V_{L2} \times \sqrt{3}I_{P2} = 3V_{L2}I_{P2} \qquad\qquad (2.144)$$

The remaining two transformers are able to supply three-phase power to the load terminal if one of the transformer is removed from the connection. This type of three-phase power supply by two transformers is known as open delta or V-V connection. This open delta connection is able to supply three-phase power at a reduced rate of 57.7 %. The connection diagram for open delta connection is shown in Fig. 2.39.

In the V-V connection, the secondary line current is,

$$I_{L2} = I_{P2} \qquad\qquad (2.145)$$

The output capacity in V-V connection is,

$$P_{0V-V} = \sqrt{3}V_{L2}I_{P2} \qquad\qquad (2.146)$$

The ratio of delta-delta output capacity to the V-V output capacity is,

$$\frac{P_{0V-V}}{P_{0\Delta-\Delta}} = \frac{\sqrt{3}V_{L2}I_{P2}}{3V_{L2}I_{P2}} = 0.577 \qquad\qquad (2.147)$$

Fig. 2.39 Open delta or V-V connection diagram

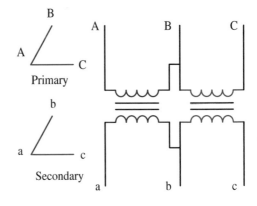

Example 2.10 A three-phase transformer is connected to an 11 kV supply and draws 6 A current. Determine (i) line voltage at the secondary side, and also (ii) the line current in the secondary coil. Consider the turns ratio of the transformer is 11. Also, consider delta-wye and wye-delta connections.

Solution
Delta-wye connection:

The phase voltage at the primary side is,

$$V_{P1} = V_{L1} = 11\,\text{kV}$$

The phase voltage at the secondary side is,

$$V_{P2} = \frac{V_{P1}}{a} = \frac{11{,}000}{11} = 1000\,\text{V}$$

(i) The line voltage at the secondary is,

$$V_{L2} = \sqrt{3}V_{P2} = \sqrt{3} \times 1000 = 1732\,\text{V}$$

(ii) The phase current in the primary is,

$$I_{P1} = \frac{I_{L1}}{\sqrt{3}} = \frac{11}{\sqrt{3}} = 6.35\,\text{A}$$

The line current in the secondary is,

$$I_{L2} = I_{P2} = aI_{P1} = 11 \times 6.35 = 69.85\,\text{A}$$

Wye-delta connection:

The line voltage at the primary side is,

$$V_{L1} = 11\,\text{kV}$$

The phase voltage at the primary side is,

$$V_{P1} = \frac{V_{L1}}{\sqrt{3}} = \frac{11{,}000}{\sqrt{3}} = 6350.85\,\text{V}$$

(i) The phase voltage at the secondary side is,

$$V_{P2} = \frac{V_{P1}}{a} = \frac{6350.85}{11} = 577.35 \text{ V}$$

The line voltage at the secondary is,

$$V_{L2} = V_{P2} = 577.35 \text{ V}$$

(ii) The phase current in the secondary is,

$$I_{P2} = aI_{P1} = 11 \times 6 = 66 \text{ A}$$

The line current in the secondary is,

$$I_{L2} = \sqrt{3}I_{P2} = \sqrt{3} \times 66 = 114.32 \text{ A}$$

Practice problem 2.10

A three-phase transformer is connected to the 33 kV supply and draws 15 A current. Calculate (i) the line voltage at the secondary side, and also (ii) the line current in the secondary winding. Consider the turns ratio of the transformer is 8 and wye-delta connection.

2.20 Instrument Transformers

The magnitude of the voltage and the current are normally high in the power system networks. To reduce the magnitude of the voltage and current, instrument transformers are used. There are two types of instrument transformers; namely, current transformer (CT) and potential transformer (PT) . If a power system carries an alternating current greater than 100 A, then it is difficult to connect measuring instruments like low range ammeter, wattmeter, energy meter and relay.

The current transformer is then connected in series with the line to step down the high magnitude of the current to a rated value of about 5 A for the ammeter and the current coil of the wattmeter. The diagram of a current transformer is shown in Fig. 2.40. If the system voltage exceeds 500 V, then it is difficult to connect measuring instruments directly to the high voltage. The potential transformer is then used to step down to a suitable value of the voltage at the secondary for supplying the voltmeter and the voltage coil of the wattmeter. The secondary of the instrument transformer is normally grounded for safety reason. The connection diagram of a potential transformer is shown in Fig. 2.41.

Fig. 2.40 Connection
diagram of current
transformer

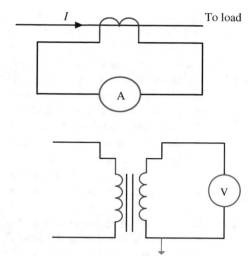

Fig. 2.41 Connection
diagram of potential
transformer

Exercise Problems

2.1   A single-phase transformer is having the primary voltage of 240 V. The
      number of turns at the primary and secondary coils are 250 and 50,
      respectively. Determine the secondary voltage.

2.2   A single-phase transformer is having the secondary voltage of 100 V. The
      number of turns at the primary and secondary windings are 500 and 50,
      respectively. Calculate the primary voltage.

2.3   The 100 A current flows in the primary coil of a single-phase transformer.
      Calculate the secondary current with the turns ratio of 0.05.

2.4   The number of turns at the primary and secondary windings of a single-phase
      transformer are 500 and 100, respectively. Find the primary current if the
      secondary current is 10 A.

2.5   The ratio of the primary current to the secondary current of a single-phase
      transformer is 1:10. Find the voltage ratio. Also, find the secondary voltage if
      the primary voltage is 100 V.

2.6   The turns ratio of a single-phase transformer is 1:4. The secondary coil has
      5000 turns and 60 V. Find the voltage ratio, primary voltage and number of
      turns.

2.7   The voltage ratio of a single-phase transformer is 5:1. The voltage and
      number of turns at the primary coil are found to be 1100 V and 500,
      respectively. Calculate the voltage and secondary number of turns at the
      secondary coil.

2.8   A single-phase transformer is connected to the 120 V source draws 4 A
      current. Calculate the current in the secondary coil when the transformer
      steps up the voltage to 500 V.

2.9 The number of turns at the primary and secondary coils of a single-phase transformer are found to be 480 and 60, respectively. The transformer draws a current of 0.6 A when connected to 120 V supply. Determine the current and the voltage at the secondary coil.

2.10 The turns ratio of a 5 kVA single-phase transformer is found to be 2. The transformer is connected to a 230 V, 50 Hz supply. If the secondary current of a transformer is 6 A, then find the primary current and secondary voltage.

2.11 The secondary number of turns of a 30 kVA, 4400/440 V single-phase transformer is found to be 100. Find the number of primary turns, full load primary and secondary currents.

2.12 A 5 kVA, 1100/230 V, 50 Hz single-phase transformer is installed near the domestic area of a country. Find the turns ratio, primary and secondary full load currents.

2.13 The number of turns of the primary winding of a single-phase 50 Hz transformer is found to be 50. Find the value of the core flux, if the induced voltage at this winding is 220 V.

2.14 The number of primary turns of a 60 Hz single-phase transformer is found to be 500. Calculate the induced voltage in this winding, if the value of the core flux is 0.05 Wb.

2.15 The maximum flux of a 3300/330 V, 50 Hz step-down single-phase transformer is found to be 0.45 Wb. Calculate the number of primary and secondary turns.

2.16 The cross sectional area of a 5 kVA, 2200/220 V, 50 Hz single-phase step-down transformer is found to be 25 cm^2 and a maximum flux density is 4 Wb/m^2. Calculate the primary and secondary turns.

2.17 The number of turns at the primary and secondary of an ideal single-phase transformer are found to be 500 and 250, respectively. The primary of the transformer is connected to a 220 V, 50 Hz source. The secondary coil supplies a current of 5 A to the load. Determine the (i) turns ratio, (ii) current in the primary and (iii) maximum flux in the core.

2.18 The primary number of turns of a 4 kVA, 1100/230 V, 50 Hz single-phase transformer is 500. The cross sectional area of the transformer core is 85 cm^2. Find the (i) turns ratio, (ii) number of turns in the secondary and (iii) maximum flux density in the core.

2.19 The primary and secondary turns of a single-phase transformer are 50 and 500, respectively. The primary winding is connected to a 220 V, 50 Hz supply. Find the (i) flux density in the core if the cross sectional area is 250 cm^2 and (ii) induced voltage at the secondary winding.

2.20 The primary coil number of turns of a single-phase 2200/220 V, 50 Hz transformer is found to be 1000. Find the area of the core if the maximum flux density of the core is 1.5 Wb/m^2.

2.21 The maximum flux of a single-phase 1100/220 V, 50 Hz transformer is found to be 5 mWb. The number of turns and the voltage at the primary winding are found to be 900 and 1100 V, respectively. Determine the frequency of the supply system.

2.22 The maximum flux density of a 6 kVA, 2200/220 V, 50 Hz single-phase transformer is found to be 0.6 Wb/m^2. The induced voltage per turn of the transformer is 10 V. Calculate the (i) primary number of turns, (ii) secondary number of turns, and (iii) cross sectional area of the core.

2.23 A single-phase transformer is having the primary voltage per turn is 20 V. Find the magnetic flux density if the cross sectional area of the core is 0.05 m^2.

2.24 The no load primary current of an 1100/230 V single-phase transformer is found to be 0.5 A and absorbs a power of 350 W from the circuit. Find the iron loss and magnetizing components of the no-load current.

2.25 A 3300 V/230 V single-phase transformer draws a no-load current of 0.5 A at a power factor of 0.4 in an open circuit test. Determine the working and magnetizing components of the no-load current.

2.26 A single-phase transformer draws a no-load current of 1.4 A from the 120 V source and absorbs a power of 80 W. The primary winding resistance and leakage reactance are found to be 0.25 and 1.2 Ω, respectively. Determine the no-load circuit resistance and reactance.

2.27 Under no-load condition, a single-phase transformer is connected to a 240 V, 50 Hz supply and draws a current of 0.3 A at a 0.5 power factor. The number of turns on the primary winding is found to be 600. Calculate the (i) maximum value of the flux in the core, (ii) magnetizing and working components of the no load current, and (iii) iron loss.

2.28 The no-load current of a 440/110 V single-phase transformer is measured 0.8 A and absorbs a power of 255 W. Calculate the (i) working and magnetizing components of the no-load current, (ii) copper loss in the primary winding if the primary winding resistance is 1.1 Ω, and (iii) core loss.

2.31 The primary winding resistance and reactance of a 300 kVA, 2200/220 V single-phase transformer are 2 and 9 Ω, respectively. The secondary winding resistance and reactance are found to be 0.02 and 0.1 Ω, respectively. Find the equivalent impedance referred to the primary and the secondary.

2.32 The primary winding resistance and reactance of a 200 kVA, 3300/220 V single-phase transformer are found to be 5 and 12 Ω, respectively. The same parameters in the secondary winding are found to 0.02 and 0.05 Ω, respectively. Find the equivalent impedance referred to the primary and the secondary. Also, determine the full load copper loss.

2.33 A 1100/400 V single-phase transformer having the resistance of 5 Ω and the reactance of 9 Ω in the primary winding. The secondary winding resistance and reactance are found to be 0.6 and 1.1 Ω respectively. Find the voltage regulation when the secondary delivers a current of 5 A at a 0.9 lagging power factor.

2.34 A 2200/220 V single-phase transformer having the resistance of 6 Ω and the reactance of 16 Ω in the primary coil. The resistance and the reactance in secondary winding are found to be 0.07 and 0.15 Ω, respectively. Find the voltage regulation when the secondary supplies a current of 6 A at a 0.86 power factor leading.

2.35 The iron loss and full load copper loss of a 40 kVA transformer are found to be 450 and 750 W, respectively. Calculate the (i) full load efficiency, (ii) output kVA corresponding to maximum efficiency, and (iii) maximum efficiency. Consider the power factor of the load is 0.95 lagging.

2.36 The iron loss of a 40 kVA transformer is 50 % of the full load copper loss. The full load copper loss is found to be 850 W. Calculate the (i) full load efficiency, (ii) output kVA corresponding to maximum efficiency, and (iii) maximum efficiency. Assume the power factor of the load is 0.9 lagging.

2.37 The iron loss and full load copper loss of a 25 kVA transformer are found to be 300 and 550 W, respectively. Find the (i) full load efficiency, (ii) output kVA corresponding to maximum efficiency, and (iii) maximum efficiency. Assume the power factor of the load is 0.6 lagging.

2.38 The primary and secondary windings parameters of a 100 kVA, 2200/220 V transformer are $R_1 = 0.6\,\Omega$, $X_1 = 0.9\,\Omega$, $R_2 = 0.007\,\Omega$, $X_2 = 0.008\,\Omega$. The transformer is operating at a maximum efficiency of 75 % of its rated load with 0.9 lagging power factor. Calculate the (i) efficiency of transformer at full load, and (ii) maximum efficiency, if the iron loss is 350 W.

2.39 A 200 kVA, 2400/240 V transformer has the primary and windings parameters $R_1 = 20\,\Omega$, $X_1 = 27\,\Omega$, $R_2 = 0.18\,\Omega$, $X_2 = 0.26\,\Omega$. The transformer is operating at a maximum efficiency of 80 % of its rated load with 0.8 pf lagging. Calculate the (i) efficiency of the transformer at full load, and (ii) maximum efficiency if the iron loss is 400 W.

2.40 The low voltage side of a 30 kVA, 2400/220 V, 50 Hz single-phase transformer is short circuited. The test data recorded from the high voltage side are 200 W, 6 A and 45 V. Find the (i) equivalent resistance, reactance and impedance referred to the primary, (ii) equivalent resistance, reactance and impedance referred to the secondary, and (iii) voltage regulation at 0.6 power factor lagging.

2.41 The low voltage winding of a 25 kVA, 1100/220 V, 50 Hz single-phase transformer is short circuited. The test data recorded from the high voltage side are 156 W, 4 A and 50 V. Calculate the (i) equivalent resistance, reactance and impedance referred to the primary, (ii) equivalent resistance, reactance and impedance referred to the secondary, and (iii) voltage regulation at 0.85 power factor leading.

2.42 The test data of a 10 kVA, 440/220 V, 50 Hz single-phase transformer are as follows:

Open circuit test: 220 V, 1.4 A, 155 W;
Short circuit test: 20.5 V, 15 A, 145 W;
Calculate the approximate equivalent circuit parameters.

2.43 The open circuit test data of a single-phase transformer are recorded as the 220 V, 1.2 A and 145 W. Find the no-load circuit parameters.

Fig. P2.1 An
autotransformer with a
specific voltage

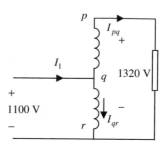

2.44 A single-phase 100 kVA, 1100/220 V transformer is connected as an auto-transformer, which is shown in Fig. P2.1. The voltage of the upper portion and lower portion of the coil are found to be 220 and 1100 V, respectively. Calculate the kVA rating of the autotransformer.

2.45 A 200 kVA three-phase, 50 Hz core-type transformer is connected as delta-wye and has a line voltage ratio of 1100/440 V. The core area and the maximum flux density of the transformer are found to be $0.04 \, m^2$ and $2.3 \, Wb/m^2$, respectively. Calculate the number of turns per phase of the primary and secondary coils.

2.46 The number of turns in the primary and secondary coils are found to be 600 and 150, respectively. The transformer is connected to 11 kV, 50 Hz supply. Find the secondary line voltage when the windings are connected as (a) delta-wye, and (b) wye-delta.

References

1. T. Wildi, *Electrical Machines, Drives and Power Systems*, 6th edn. (Pearson Education International, Upper Saddle River, 2006)
2. M.A. Salam, *Fundamentals of Electrical Machines*, 2nd edn. (Alpha Science, International Ltd., Oxford, 2012)
3. S.J. Chapman, *Electric Machinery and Power System Fundamentals* (McGraw-Hill Higher Education, New York, 2002)
4. A.E. Fitzgerald, C. Kingsley Jr., S.D. Umans, Electric Machinery, 6th edn (McGraw-Hill Higher Education, New York, 2003)

Chapter 3
Symmetrical and Unsymmetrical Faults

3.1 Introduction

The goal of any power utility company is to run its power system network under balanced condition. The power system network is said to be balanced when it is operating in normal-load condition. This normal operating condition of the power system can be disrupted due to adverse weather condition such as heavy wind, lightning strikes etc., or due to other factors such as birds shorting out the lines, vehicles collides with transmission-line poles or towers by accident or trees fall down on the transmission lines. The lightning strikes on the transmission line may generate very high transient voltage which exceeds the basic insulation voltage level of the transmission lines. This event triggers the flashover from the resultant high magnitude current that passes through the transmission-tower to the ground. This condition of the transmission lines is known as a short circuit condition, and the fault associated to this phenomenon is known as short circuit fault. In a short circuit situation, a very low impedance-path is created either in between two transmission lines or in between a transmission line and ground. In this case, the resulting high magnitude current imposes heavy duty on the circuit breaker and other controlling equipment. The short circuit faults are classified as symmetrical and unsymmetrical faults. In this chapter, symmetrical and unsymmetrical faults, symmetrical components, zero sequence components of the machines and classification of unsymmetrical faults will be discussed.

3.2 Symmetrical Faults

The symmetrical faults are often known as balanced faults. In case of balanced faults, all three transmission lines are affected equally, and the system remains in the balanced condition. These types of faults are rare in the power system, and it

© Springer Science+Business Media Singapore 2016
Md.A. Salam and Q.M. Rahman, *Power Systems Grounding*,
Power Systems, DOI 10.1007/978-981-10-0446-9_3

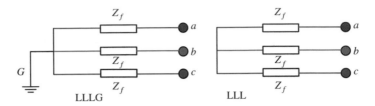

Fig. 3.1 Connection diagrams of symmetrical faults

contributes 2 % to 5 % of the total fault. These faults are easy to analyze. The symmetrical faults are classified as three lines to ground fault (LLLG) and three lines fault (LLL). The connection diagrams of symmetrical faults are shown in Fig. 3.1. If the fault impedance, $Z_f = 0$, then the fault is known as solid or bolted fault.

3.3 Unsymmetrical Faults

The faults in the power system network which disturb the balanced condition of the network are known as unsymmetrical faults. The unsymmetrical faults are classified as single line to ground faults (SLG), double line to ground faults (DLG) and line to line faults (LL). More than 90 % faults occur in a power system are single line to ground faults. The connection diagrams of different types of unsymmetrical faults are shown in Fig. 3.2.

3.4 Symmetrical Components

The knowledge of phase sequence is very important to analyze the symmetrical components. The phase sequence is defined as the order in which they reach to the maximum value of the voltage. In 1918, C. L. Fortescue, an American Scientist stated

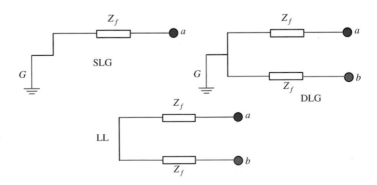

Fig. 3.2 Connection diagrams of unsymmetrical faults

that the three-phase unbalanced phasors of a three-phase system can be replaced by three separate balanced phasors. The symmetrical components are classified as positive sequence components, negative sequence components, and zero sequence components. The components of the system have three phasors with equal magnitude but are displaced from each other by 120° and they follow the *abc* phase sequence.

The phase *b* lags the phase *a* by 120° and the phase *c* lags the phase *b* by 120°. The neutral current of the positive sequence components is zero. The positive sequence components I_{a1}, I_{b1} and I_{c1} are shown in Fig. 3.3. In this system, the components have three phasors with equal magnitude but are displaced from each other by 120° and they maintain the *acb* phase sequence. The phase *c* lags the phase *a* by 120° and the phase *b* lags the phase *c* by 120°. The negative sequence components are similar to the positive sequence components, except the phase order is reversed. The neutral current of the negative sequence components is zero. The negative sequence components I_{a2}, I_{b2} and I_{c2} are shown in Fig. 3.4.

The components have three phasors with equal magnitude but zero displacement are known as zero sequence components. The zero sequence components are in phase with each other and have the neutral current. The zero sequence components of the current I_{a0}, I_{b0} and I_{c0} are shown in Fig. 3.5.

The unsymmetrical components of the currents I_a, I_b and I_c can be derived from the phasor diagram as shown in Fig. 3.6.

Fig. 3.3 Positive sequence components

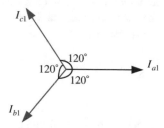

Fig. 3.4 Negative sequence components

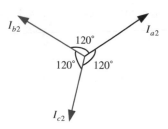

Fig. 3.5 Zero sequence components

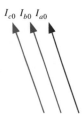

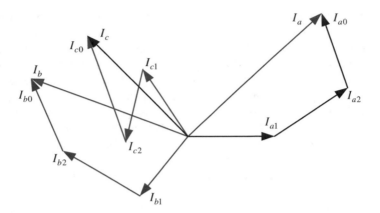

Fig. 3.6 Unsymmetrical and symmetrical components of current

3.5 Representation of Symmetrical Components

Figure 3.7 represents a phasor diagram where the line *OA* represents the current
phasor *I*. This phasor, after being multiplied by the operator *a* gives the new phasor
aI which is represented by the line *OB*. The phasor *aI* in the diagram is leading (in
the counterclockwise direction) the phasor *I* by 120°, which can be mathematically
expressed as,

$$aI = I \underline{|120°}$$ (3.1)

$$a = 1 \underline{|120°} = -0.5 + j0.866$$ (3.2)

Similarly, multiplying the phasor *aI* by *a* which gives the third phasor $a^2 I$, which
is represented by the line *OC*. This new phasor $a^2 I$ is leading phasor *I* by 240° in
the phasor diagram, which can be mathematically expressed as,

Fig. 3.7 Phasors with
counterclockwise direction

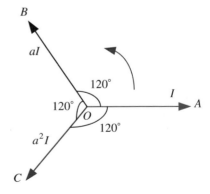

$$a^2 I = I \underline{|240°}$$ (3.3)

$$a^2 = 1 \underline{|240°} = -0.5 - j0.866$$ (3.4)

A similar approach can lead us to the following expression:

$$a^3 = 1 \underline{|360°} = 1$$ (3.5)

Adding Eqs. (3.2), (3.4) and (3.5) yields,

$$1 + a^2 + a^3 = 0$$ (3.6)

Comparing Fig. 3.3 with Fig. 3.7 the positive phase sequence components of the current can be represented as,

$$I_{a1} = I_{a1} \underline{|0°}$$ (3.7)

$$I_{b1} = I_{a1} \underline{|240°} = a^2 I_{a1}$$ (3.8)

$$I_{c1} = I_{a1} \underline{|120°} = a I_{a1}$$ (3.9)

Again, comparing Fig. 3.4 with Fig. 3.7 the negative phase sequence components of the current can be represented as,

$$I_{a2} = I_{a2} \underline{|0°}$$ (3.10)

$$I_{b2} = I_{a2} \underline{|120°} = a I_{a2}$$ (3.11)

$$I_{c2} = I_{a2} \underline{|240°} = a^2 I_{a2}$$ (3.12)

The magnitudes of the zero phase sequence components are the same, and it can be written as,

$$I_{a0} = I_{b0} = I_{c0}$$ (3.13)

From Fig. 3.6, the unsymmetrical currents can be represented by the symmetrical components of current as,

$$I_a = I_{a0} + I_{a1} + I_{a2}$$ (3.14)

$$I_b = I_{b0} + I_{b1} + I_{b2}$$ (3.15)

$$I_c = I_{c0} + I_{c1} + I_{c2}$$ (3.16)

where the suffixes 0, 1 and 2 indicates the zero sequence, positive sequence and negative sequence components, respectively. Equations (3.15) and (3.16) can be replaced by the symmetrical components of the current of phase a. Substituting Eqs. (3.8), (3.11) and (3.13) into Eq. (3.15) yields,

$$I_b = I_{a0} + a^2 I_{a1} + a I_{a2} \qquad (3.17)$$

Again, substituting Eqs. (3.9), (3.12) and (3.13) into the Eq. (3.16) yields,

$$I_c = I_{a0} + a I_{a1} + a^2 I_{a2} \qquad (3.18)$$

Equations (3.14), (3.17) and (3.18) can be re-arranged in the matrix form as,

$$\begin{bmatrix} I_a \\ I_b \\ I_c \end{bmatrix} = \begin{bmatrix} 1 & 1 & 1 \\ 1 & a^2 & a \\ 1 & a & a^2 \end{bmatrix} \begin{bmatrix} I_{a0} \\ I_{a1} \\ I_{a2} \end{bmatrix} \qquad (3.19)$$

In Eq. (3.19), let's represent the following symmetrical component transformation matrix as A where,

$$A = \begin{bmatrix} 1 & 1 & 1 \\ 1 & a^2 & a \\ 1 & a & a^2 \end{bmatrix} \qquad (3.20)$$

Equation (3.19) can be modified as,

$$\begin{bmatrix} I_a \\ I_b \\ I_c \end{bmatrix} = A \begin{bmatrix} I_{a0} \\ I_{a1} \\ I_{a2} \end{bmatrix} \qquad (3.21)$$

Taking the inverse of A, Eq. (3.21) can be written as,

$$\begin{bmatrix} I_{a0} \\ I_{a1} \\ I_{a2} \end{bmatrix} = A^{-1} \begin{bmatrix} I_a \\ I_b \\ I_c \end{bmatrix} \qquad (3.22)$$

From Eq. (3.22), the inverse of the A can be derived as,

$$A^{-1} = \frac{1}{3} \begin{bmatrix} 1 & 1 & 1 \\ 1 & a & a^2 \\ 1 & a^2 & a \end{bmatrix} \qquad (3.23)$$

Substituting Eq. (3.23) into Eq. (3.22) yields,

$$\begin{bmatrix} I_{a0} \\ I_{a1} \\ I_{a2} \end{bmatrix} = \frac{1}{3} \begin{bmatrix} 1 & 1 & 1 \\ 1 & a & a^2 \\ 1 & a^2 & a \end{bmatrix} \begin{bmatrix} I_a \\ I_b \\ I_c \end{bmatrix} \qquad (3.24)$$

From Eq. (3.24), the following expressions for the current components can be found,

$$I_{a0} = \frac{1}{3}(I_a + I_b + I_c) \tag{3.25}$$

$$I_{a1} = \frac{1}{3}(I_a + aI_b + a^2 I_c) \tag{3.26}$$

$$I_{a2} = \frac{1}{3}(I_a + a^2 I_b + aI_c) \tag{3.27}$$

Similarly, the mathematical expressions for voltage components can be written as,

$$V_{a0} = \frac{1}{3}(V_a + V_b + V_c) \tag{3.28}$$

$$V_{a1} = \frac{1}{3}(V_a + aV_b + a^2 V_c) \tag{3.29}$$

$$V_{a2} = \frac{1}{3}(V_a + a^2 V_b + aV_c) \tag{3.30}$$

In a three-phase Y-connection system, the magnitude of the neutral current can be calculated as,

$$I_n = I_a + I_b + I_c \tag{3.31}$$

Substituting Eq. (3.31) into Eq. (3.25) yields,

$$I_{a0} = \frac{1}{3}I_n \tag{3.32}$$

$$I_n = 3I_{a0} = 3I_0 \tag{3.33}$$

Example 3.1 A three-phase system is having phase voltages are $V_a = 90\underline{|0°}$ kV, $V_b = 66\underline{|-100°}$ kV, and $V_c = 22\underline{|85°}$ kV. Find the symmetrical voltage components of for phases a, b and c.

Solution

The magnitude of the zero sequence voltage component is,

$$V_{a0} = \frac{1}{3}(90 + 66\underline{|-100°} + 22\underline{|85°})$$
$$V_{a0} = 30.42\underline{|-28.17°} \text{ kV}$$

The magnitude of the positive sequence voltage component is,

$$V_{a1} = \frac{1}{3}(90 + 66\angle{-100° + 120°} + 22\angle{85° + 240°})$$
$$V_{a1} = \frac{1}{3}(90 + 66\angle{20°} + 22\angle{325°})$$
$$V_{a1} = 56.21\angle{3.35°}\ \text{kV}$$

The magnitude of the negative sequence voltage component is,

$$V_{a2} = \frac{1}{3}(V_a + a^2 V_b + a V_c) = \frac{1}{3}(90 + 66\angle{-100° + 240°} + 22\angle{85° + 120°})$$
$$V_{a2} = \frac{1}{3}(90 + 66\angle{140°} + 22\angle{205°})$$
$$V_{a2} = 12.69\angle{59.51°}\ \text{kV}$$

For phase b:
The zero sequence voltage component is,

$$V_{b0} = 30.42\angle{-28.17°}\ \text{kV}$$

The positive sequence voltage component is,

$$V_{b1} = a^2 V_{a1} = 56.21\angle{240° + 3.35°} = 56.21\angle{243.35°}\ \text{kV}$$

The negative sequence voltage component is calculated as,

$$V_{b2} = V_{a2}\angle{120°} = 12.69\angle{120° + 59.51°} = 12.69\angle{179.51°}\ \text{kV}$$

For phase c:
The zero sequence voltage component is,

$$V_{c0} = 30.42\angle{-28.17°}\ \text{kV}$$

The positive sequence voltage component can be determined as,

$$V_{c1} = V_{a1}\angle{120°} = 56.21\angle{120° + 3.35°} = 56.21\angle{123.35°}\ \text{kV}$$

The negative sequence voltage component can be calculated is,

$$V_{c2} = V_{a2}\angle{240°} = 12.69\angle{240° + 59.51°} = 12.69\angle{299.51°}\ \text{kV}$$

Example 3.2 A three-phase system is having the symmetrical voltage components of $V_{a0} = 94\angle{150°}$ kV, $V_{a1} = 56\angle{80°}$ kV, and $V_{a2} = 122\angle{55°}$ kV for phase a. Find the three-phase unbalanced voltages.

Solution

The positive phase sequence voltage components can be determined as,

$$V_{b1} = V_{a1}\underline{|240°} = 56\underline{|80 + 240°} = 56\underline{|320°}\ \text{kV}$$
$$V_{c1} = V_{a1}\underline{|120°} = 56\underline{|80° + 120°} = 56\underline{|200°}\ \text{kV}$$

The negative phase sequence voltage components are calculated as,

$$V_{b2} = V_{a2}\underline{|120°} = 122\underline{|55° + 120°} = 122\underline{|175°}\ \text{kV}$$
$$V_{c2} = V_{a2}\underline{|240°} = 122\underline{|55° + 240°} = 122\underline{|295°}\ \text{kV}$$

The zero phase sequence voltage components can be calculated as,

$$V_{a0} = V_{b0} = V_{c0} = 94\underline{|150°}\ \text{kV}$$

Unbalance three-phase voltage components can be determined as,

$$V_a = 94\underline{|150°} + 56\underline{|80°} + 122\underline{|55°} = 202.09\underline{|90.48°}\ \text{kV}$$
$$V_a = 94\underline{|150°} + 56\underline{|320°} + 122\underline{|175°} = 161.50\underline{|172.5°}\ \text{kV}$$
$$V_c = 94\underline{|150°} + 56\underline{|200°} + 122\underline{|295°} = 116.81\underline{|-134.91°}\ \text{kV}$$

Practice problem 3.1
A Y-connected three-phase load draws current from the source in the amount $I_a = 15\underline{|0°}$ A, $I_b = 25\underline{|20°}$ A, and $I_c = 30\underline{|40°}$ A. Calculate the symmetrical current components for the phases a, b and c.

Practice problem 3.2
A three-phase system has the following symmetrical current components: Positive sequence current components are $I_{a1} = 6\underline{|10°}$ A, and $I_{b1} = 6\underline{|250°}$ A. The negative sequence current components are $I_{a2} = 5\underline{|53.13°}$ A, $I_{b2} = 5\underline{|173.13°}$ A and $I_{c2} = 5\underline{|293.13°}$ A. The zero sequence current components are $I_{a0} = I_{b0} = I_{c0} = 4\underline{|35°}$ A. Calculate the unsymmetrical current components.

3.6 Complex Power in Symmetrical Components

The product of the voltage and the conjugate of the current is known as complex power and it is denoted by the capital letter S. The expression of the complex power is,

$$\mathbf{S} = P + jQ = VI^* = VI\cos\phi + VI\sin\phi \tag{3.34}$$

Based on Eq. (3.34), the expression of the complex power for the three-phase lines can be written as,

$$P + jQ = V_a I_a^* + V_b I_b^* + V_c I_c^* \tag{3.35}$$

Equation (3.35) can be re-arranged in the matrix form as,

$$P + jQ = [V_a + V_b + V_c] \begin{bmatrix} I_a \\ I_b \\ I_c \end{bmatrix}^* = \begin{bmatrix} V_a \\ V_b \\ V_c \end{bmatrix}^T \begin{bmatrix} I_a \\ I_b \\ I_c \end{bmatrix}^* \tag{3.36}$$

According to Eq. (3.19), the unbalanced voltages can be written as,

$$\begin{bmatrix} V_a \\ V_b \\ V_c \end{bmatrix} = \begin{bmatrix} 1 & 1 & 1 \\ 1 & a^2 & a \\ 1 & a & a^2 \end{bmatrix} \begin{bmatrix} V_{a0} \\ V_{a1} \\ V_{a2} \end{bmatrix} = A V_{012} \tag{3.37}$$

where,

$$\begin{bmatrix} V_{a0} \\ V_{a1} \\ V_{a2} \end{bmatrix} = V_{012} \tag{3.38}$$

Taking transpose of Eq. (3.37) yields,

$$\begin{bmatrix} V_a \\ V_b \\ V_c \end{bmatrix}^T = (A V_{012})^T = A^T V_{012}^T \tag{3.39}$$

$$\begin{bmatrix} V_a \\ V_b \\ V_c \end{bmatrix}^T = \begin{bmatrix} 1 & 1 & 1 \\ 1 & a^2 & a \\ 1 & a & a^2 \end{bmatrix} [V_{a0} \quad V_{a1} \quad V_{a2}] \tag{3.40}$$

Taking conjugate of Eq. (3.19) yields,

$$\begin{bmatrix} I_a \\ I_b \\ I_c \end{bmatrix}^* = \begin{bmatrix} 1 & 1 & 1 \\ 1 & a^2 & a \\ 1 & a & a^2 \end{bmatrix}^* \begin{bmatrix} I_{a0} \\ I_{a1} \\ I_{a2} \end{bmatrix}^* = \begin{bmatrix} 1 & 1 & 1 \\ 1 & a & a^2 \\ 1 & a^2 & a \end{bmatrix} \begin{bmatrix} I_{a0} \\ I_{a1} \\ I_{a2} \end{bmatrix}^* \tag{3.41}$$

Substituting Eqs. (3.40) and (3.41) into Eq. (3.36) yields,

$$P + jQ = \begin{bmatrix} 1 & 1 & 1 \\ 1 & a^2 & a \\ 1 & a & a^2 \end{bmatrix} [V_{a0} \quad V_{b0} \quad V_{c0}] \begin{bmatrix} 1 & 1 & 1 \\ 1 & a & a^2 \\ 1 & a^2 & a \end{bmatrix} \begin{bmatrix} I_{a0} \\ I_{a1} \\ I_{a2} \end{bmatrix}^* \tag{3.42}$$

$$P + jQ = \begin{bmatrix} V_{a0} & V_{a1} & V_{a2} \end{bmatrix} \begin{bmatrix} 1 & 1 & 1 \\ 1 & a^2 & a \\ 1 & a & a^2 \end{bmatrix} \begin{bmatrix} 1 & 1 & 1 \\ 1 & a & a^2 \\ 1 & a^2 & a \end{bmatrix} \begin{bmatrix} I_{a0} \\ I_{a1} \\ I_{a2} \end{bmatrix}^* \qquad (3.43)$$

Equation (3.43) can be modified as,

$$P + jQ = \begin{bmatrix} V_{a0} & V_{a1} & V_{a2} \end{bmatrix} \begin{bmatrix} 1+1+1 & 1+a+a^2 & 1+a^2+a \\ 1+a^2+a & 1+a^3+a^3 & 1+a^4+a^2 \\ 1+a+a^2 & 1+a^2+a^4 & 1+a^3+a^3 \end{bmatrix} \begin{bmatrix} I_{a0} \\ I_{a1} \\ I_{a2} \end{bmatrix}^* \qquad (3.44)$$

$$P + jQ = \begin{bmatrix} V_{a0} & V_{a1} & V_{a2} \end{bmatrix} \begin{bmatrix} 3 & 0 & 0 \\ 0 & 3 & 0 \\ 0 & 0 & 3 \end{bmatrix} \begin{bmatrix} I_{a0} \\ I_{a1} \\ I_{a2} \end{bmatrix}^* \qquad (3.45)$$

$$P + jQ = 3 \begin{bmatrix} V_{a0} & V_{a1} & V_{a2} \end{bmatrix} \begin{bmatrix} I_{a0} \\ I_{a1} \\ I_{a2} \end{bmatrix}^* \qquad (3.46)$$

$$P + jQ = 3 \begin{bmatrix} V_{a0} I_{a0}^* + V_{a1} I_{a1}^* + V_{a2} I_{a2}^* \end{bmatrix} \qquad (3.47)$$

Equation (3.47) can be used to find the real power and reactive power from the symmetrical components of voltage and current. According to Eq. (3.34), the expressions of real and reactive power from Eq. (3.47) can be written as,

$$P = 3V_{a0} I_{a0} \cos \phi_0 + 3V_{a1} I_{a1} \cos \phi_1 + 3V_{a2} I_{a2} \cos \phi_2 \qquad (3.48)$$

$$Q = 3V_{a0} I_{a0} \sin \phi_0 + 3V_{a1} I_{a1} \sin \phi_1 + 3V_{a2} I_{a2} \sin \phi_2 \qquad (3.49)$$

3.7 Sequence Impedances of Power System Equipment

The sequence impedances of a power system equipment is defined as the impedance offered by the equipment to the flow of sequence (positive or negative or zero) current through it. These sequence impedances are zero, positive and negative sequence impedances. The zero sequence impedance of an equipment is defined as the impedance offered by the equipment to the flow of the zero sequence current and it is represented by Z_0. The impedance offered by the power system equipment to the flow of the positive sequence current is known as the positive sequence impedance and it is denoted by Z_1. The impedance offered by the power system equipment to the flow of the negative sequence current is known as the negative sequence impedance and it is denoted by Z_2.

Fig. 3.8 A Y-connected
balanced load

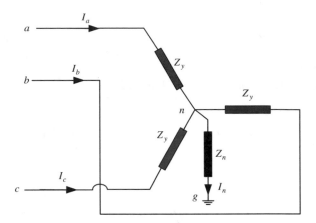

In case of synchronous machine, positive sequence impedance is equal to the synchronous impedance of the machine. Whereas, the negative sequence impedance is much less than the positive sequence impedance. If the zero sequence impedance is not given then its value is assumed to be equal to the positive sequence impedance. For transformer, positive sequence impedance, negative sequence impedance and zero sequence impedance are equal. In case of transmission line, positive sequence impedance and negative sequence impedance are equal. The zero sequence impedance is much higher than the positive sequence impedance or the negative sequence impedance.

The balanced Y-connected load and the neutral impedance is shown in Fig. 3.8. The current in the neutral point is,

$$I_n = I_a + I_b + I_c \tag{3.50}$$

The voltage between the phase a and the ground point is,

$$V_{ag} = I_a Z_y + Z_n I_n \tag{3.51}$$

Substituting Eq. (3.50) into Eq. (3.51) yields,

$$V_{ag} = I_a Z_y + Z_n (I_a + I_b + I_c) \tag{3.52}$$

$$V_{ag} = I_a (Z_y + Z_n) + Z_n I_b + Z_n I_c \tag{3.53}$$

Similarly, the voltages of phase b and phase c to the ground point are,

$$V_{bg} = Z_n I_a + (Z_n + Z_y) I_b + Z_n I_c \tag{3.54}$$

$$V_{cg} = Z_n I_a + Z_n I_b + (Z_n + Z_y) I_c \tag{3.55}$$

Equations (3.53), (3.54) and (3.55) can be expressed in the matrix format as,

$$
\begin{bmatrix} V_{ag} \\ V_{bg} \\ V_{cg} \end{bmatrix} = \begin{bmatrix} Z_n + Z_y & Z_n & Z_n \\ Z_n & Z_n + Z_y & Z_n \\ Z_n & Z_n & Z_n + Z_y \end{bmatrix} \begin{bmatrix} I_a \\ I_b \\ I_c \end{bmatrix}
\tag{3.56}
$$

Substituting the expression of unsymmetrical voltages into Eq. (3.56) yields,

$$
\begin{bmatrix} 1 & 1 & 1 \\ 1 & a^2 & a \\ 1 & a & a^2 \end{bmatrix} \begin{bmatrix} V_{a0} \\ V_{a1} \\ V_{a2} \end{bmatrix} = \begin{bmatrix} Z_n + Z_y & Z_n & Z_n \\ Z_n & Z_n + Z_y & Z_n \\ Z_n & Z_n & Z_n + Z_y \end{bmatrix} \begin{bmatrix} I_a \\ I_b \\ I_c \end{bmatrix}
\tag{3.57}
$$

Again, substituting the expression of unsymmetrical currents into Eq. (3.57) yields,

$$
\begin{bmatrix} V_{a0} \\ V_{a1} \\ V_{a2} \end{bmatrix} = \frac{1}{3} \begin{bmatrix} 1 & 1 & 1 \\ 1 & a & a^2 \\ 1 & a^2 & a \end{bmatrix} \begin{bmatrix} Z_n + Z_y & Z_n & Z_n \\ Z_n & Z_n + Z_y & Z_n \\ Z_n & Z_n & Z_n + Z_y \end{bmatrix} \begin{bmatrix} 1 & 1 & 1 \\ 1 & a^2 & a \\ 1 & a & a^2 \end{bmatrix} \begin{bmatrix} I_{a0} \\ I_{a1} \\ I_{a2} \end{bmatrix}
\tag{3.58}
$$

$$
\begin{bmatrix} V_{a0} \\ V_{a1} \\ V_{a2} \end{bmatrix} = \frac{1}{3} \begin{bmatrix} 1 & 1 & 1 \\ 1 & a & a^2 \\ 1 & a^2 & a \end{bmatrix} \begin{bmatrix} 3Z_n + Z_y & Z_n + Z_y + Z_n(a^2 + a) & Z_n + Z_y + Z_n(a^2 + a) \\ 3Z_n + Z_y & Z_n + a^2(Z_y + Z_n) + aZ_n & Z_n + a(Z_y + Z_n) + a^2 Z_n \\ 3Z_n + Z_y & Z_n + a^2 Z_n + a(Z_n + Z_y) & Z_n + aZ_n + a^2(Z_y + Z_n) \end{bmatrix} \begin{bmatrix} I_{a0} \\ I_{a1} \\ I_{a2} \end{bmatrix}
\tag{3.59}
$$

$$
\begin{bmatrix} V_{a0} \\ V_{a1} \\ V_{a2} \end{bmatrix} = \frac{1}{3} \begin{bmatrix} 1 & 1 & 1 \\ 1 & a & a^2 \\ 1 & a^2 & a \end{bmatrix} \begin{bmatrix} 3Z_n + Z_y & Z_y & Z_y \\ 3Z_n + Z_y & a^2 Z_y & aZ_y \\ 3Z_n + Z_y & aZ_y & a^2 Z_y \end{bmatrix} \begin{bmatrix} I_{a0} \\ I_{a1} \\ I_{a2} \end{bmatrix}
\tag{3.60}
$$

$$
\begin{bmatrix} V_{a0} \\ V_{a1} \\ V_{a2} \end{bmatrix} = \frac{1}{3} \begin{bmatrix} 9Z_n + 3Z_y & Z_y(a^2 + a + 1) & Z_y(a^2 + a + 1) \\ (3Z_n + Z_y)(a^2 + a + 1) & Z_y + a^3 Z_y + a^3 Z_y & Z_y + a^2 Z_y + a^4 Z_y \\ (3Z_n + Z_y)(a^2 + a + 1) & Z_y + a^4 Z_y + a^2 Z_y & Z_y + a^3 Z_y + a^3 Z_y \end{bmatrix} \begin{bmatrix} I_{a0} \\ I_{a1} \\ I_{a2} \end{bmatrix}
\tag{3.61}
$$

$$
\begin{bmatrix} V_{a0} \\ V_{a1} \\ V_{a2} \end{bmatrix} = \frac{1}{3} \begin{bmatrix} 9Z_n + 3Z_y & 0 & 0 \\ 0 & 3Z_y & 0 \\ 0 & 0 & 3Z_y \end{bmatrix} \begin{bmatrix} I_{a0} \\ I_{a1} \\ I_{a2} \end{bmatrix}
\tag{3.62}
$$

$$
\begin{bmatrix} V_{a0} \\ V_{a1} \\ V_{a2} \end{bmatrix} = \begin{bmatrix} 3Z_n + Z_y & 0 & 0 \\ 0 & Z_y & 0 \\ 0 & 0 & Z_y \end{bmatrix} \begin{bmatrix} I_{a0} \\ I_{a1} \\ I_{a2} \end{bmatrix}
\tag{3.63}
$$

Fig. 3.9 Sequence networks

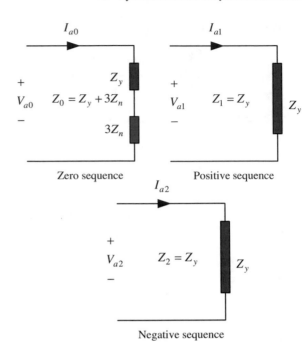

From Eq. (3.63), the expressions of symmetrical components of voltage for the phase a can be written as,

$$V_{a0} = (3Z_n + Z_y)I_{a0} \qquad (3.64)$$

$$V_{a1} = Z_y I_{a1} \qquad (3.65)$$

$$V_{a2} = Z_y I_{a2} \qquad (3.66)$$

The sequence circuits based on Eqs. (3.64), (3.65) and (3.66) are shown in Fig. 3.9. The neutral impedance of the Y-connection would be zero if it is solidly grounded. Equation (3.64) can be modified as,

$$V_{a0} = Z_y I_{a0} \qquad (3.67)$$

The neutral impedance of the Y-connection will provide an infinite quantity if it is not grounded. That means zero sequence circuit will provide an open circuit and non zero sequence current will flow through an ungrounded Y-connection.

3.8   Zero Sequence Models

In transmission lines, there is no effect on the impedance due to positive and native sequence components of voltages and currents. In this case, positive sequence and negative sequence impedances are equal to each other, that is,

$$Z_1 = Z_2 \tag{3.68}$$

The zero sequence impedance is much higher than the positive or negative sequence impedance due to its ground return path. In this case, the expression can be written as,

$$Z_0 = Z_1 + 3Z_n \tag{3.69}$$

Similarly, the zero sequence reactance can be written as,

$$X_0 = X_1 + 3X_n \tag{3.70}$$

where the neutral reactance (per mile) can be expressed as,

$$X_n = 2.02 \times 10^{-3} f \ln \frac{D_m}{D_s} \ \Omega/\text{mile} \tag{3.71}$$

where,

D_m is the geometric mean distance (GMD),
D_s is the geometric mean radius,
f is the frequency.

The sequence networks for transmission line are shown in Fig. 3.10.

There are three impedances namely sub-transient, transient and direct axis reactance. The positive and negative sequence impedances are equal to the

Fig. 3.10 Sequence networks for transmission line

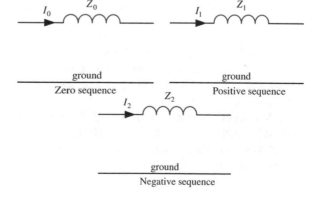

Fig. 3.11 Sequence networks for generator

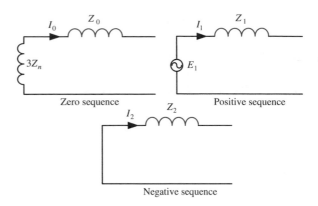

sub-transient reactance during fault condition. A generator in a power system produces positive sequence voltage only. Therefore, voltage source is present in the positive sequence. The generator offers a very small reactance due to the leakage flux. Therefore, zero sequence impedance is smaller than others. The following equations for a generator can be written as,

$$Z_1 = Z_2 = Z_d'' \tag{3.72}$$

$$Z_0 = Z_l \tag{3.73}$$

The sequence networks for the generator is shown in Fig. 3.11.

In the transformer, the zero sequence current flows if the neutral is grounded. In this case, the positive and the negative sequence impedance is equal to the zero sequence impedance, that is,

$$Z_1 = Z_2 = Z_0 \tag{3.74}$$

Y-Y Connection with both neutral grounded: In this connection, the both neutrals are connected to the ground. Therefore, zero sequence current will flow in the primary and the secondary windings through the two grounded neutrals. As a result, the zero sequence impedance connects the high voltage and the low voltage terminals as shown in Fig. 3.12.

Y-Y connection with only one grounded: In this arrangement, one side of the Y-connection is not grounded. If any one of the Y-Y connection is not grounded then the zero sequence current will not flow through the ungrounded wye. Therefore, an open circuit will appear between the high voltage and low voltage side as shown in Fig. 3.13.

Y-Y connection with no grounded: In this arrangement, the sum of the phase current in both cases is zero. As a result, no zero sequence current will flow in any of the windings. Hence, there is an open circuit between the high voltage and the low voltage sides as shown in Fig. 3.14.

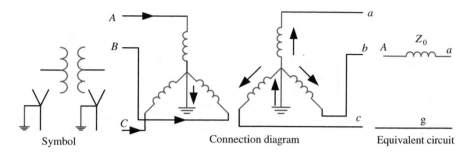

Fig. 3.12 Sequence networks for Y-Y connection with both grounded

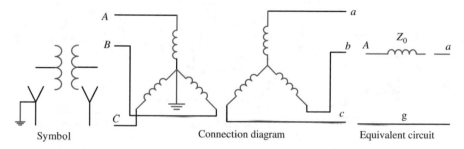

Fig. 3.13 Sequence networks for Y-Y connection with one grounded

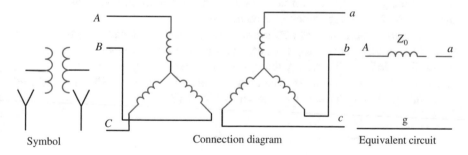

Fig. 3.14 Sequence networks for Y-Y connection with no grounded

Grounded Y and Δ connections: In this connection, zero sequence current will flow in the Y-connection as its neutral is connected to the ground. The balanced zero sequence current will flow within the phases of the Δ connection but this current will not flow out of the terminal. Therefore, no zero sequence current will flow in the line as shown in Fig. 3.15.

Ungrounded Y and Δ connections: In this case, there will be no connection between the neutral and the ground. Therefore, zero sequence current will not flow

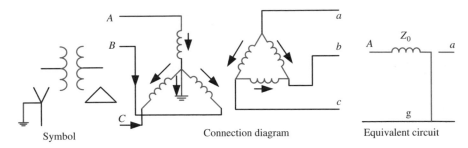

Fig. 3.15 Sequence networks for grounded Y and delta connection

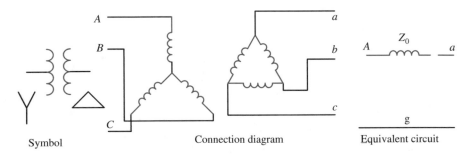

Fig. 3.16 Sequence networks for ungrounded Y and delta connection

in the windings of both transformers. As a result, an open circuit will exists between the high and low voltage sides as shown in Fig. 3.16.

Δ–Δ **connection**: In this connection, no zero sequence current will leave or enter the terminals. However, it is possible for the current components to circulate within the windings. Therefore, there is an open circuit between the high voltage and the low voltage windings. The zero sequence impedance will form a closed path with grounding terminals as shown in Fig. 3.17.

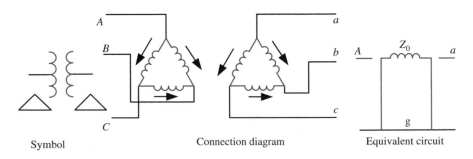

Fig. 3.17 Sequence networks for delta-delta connection

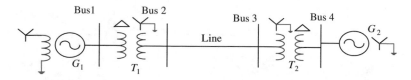

Fig. 3.18 A single-line diagram of a power system

Fig. 3.19 A positive sequence network

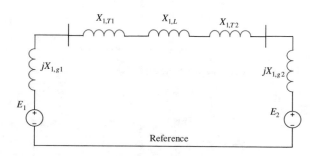

Fig. 3.20 A negative sequence network

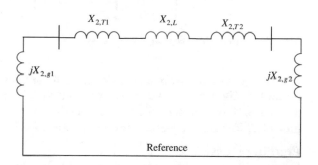

Example 3.3 Figure 3.18 shows a single-line diagram of a three-phase power system. Draw the positive sequence, negative sequence and zero sequence networks.

Solution

Generator is represented with a voltage source and a series reactance in the positive sequence network. Transformer and transmission line are also represented by the respective reactance quantities as shown in Fig. 3.19.

By omitting the voltage sources from the positive sequence network and by replacing the generator reactance components with negative sequence reactance components as shown in Fig. 3.20 has been derived.

The equivalent reactance of the generator G_1 is $X_{g0} + 3X_n$ as it is grounded through a reactance. The total equivalent reactance of the generator G_2 is zero as it

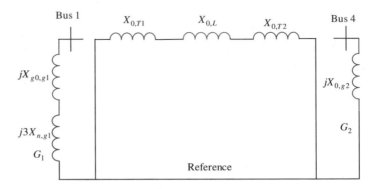

Fig. 3.21 A zero sequence network

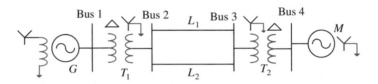

Fig. 3.22 A single-line diagram of a power system

is solidly grounded. The primary sides of both the transformers T_1 and T_2 are delta connected. The zero sequence network is therefore, in an open circuit near the bus 1 and bus 4. The secondary of both the transformers are wye-connected and solidly grounded. The zero sequence network is shown in Fig. 3.21.

Practice problem 3.3
Figure 3.22 shows a single-line diagram of a three-phase power system. Draw the positive sequence, negative sequence and zero sequence networks.

3.9   Classification of Unsymmetrical Faults

Unsymmetrical faults are the most common faults that occur in the power system. Due to this fault, the magnitudes of the line currents becomes unequal, and also these current components observed a phase displacement among them. In this case, symmetrical components are required to analyze the current and voltage quantities during the unsymmetrical faults. These unsymmetrical faults can be classified into three categories, namely, single line-to-ground fault (SLG), line-to-line fault (LL) and double line-to-ground fault (DLG). The unsymmetrical faults are shown in Fig. 3.23.

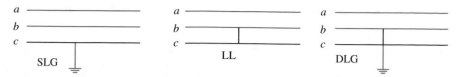

Fig. 3.23 Different types of unsymmetrical faults

3.10 Sequence Network of an Unloaded Synchronous Generator

A three-phase unloaded synchronous generator is having a synchronous impedance of Z_y per phase and its neutral is grounded by an impedance Z_n as shown in Fig. 3.24. In balanced condition, the negative sequence and zero sequence voltages are zero. The expression of neutral current is [1–6],

$$I_n = I_a + I_b + I_c \tag{3.75}$$

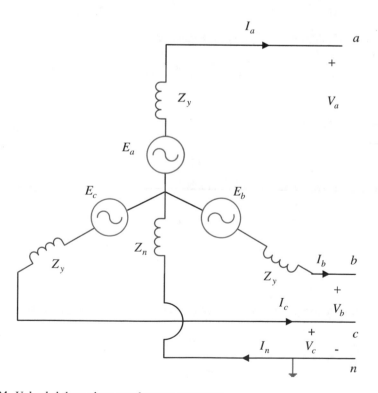

Fig. 3.24 Unloaded three-phase synchronous generator

Applying KVL to the circuit shown in Fig. 3.24, the following equation can be found:

$$V_a = E_a - Z_y I_a - Z_n I_n \tag{3.76}$$

Substituting Eq. (3.75) into Eq. (3.76) yields,

$$E_a = V_a + Z_y I_a + Z_n (I_a + I_b + I_c) \tag{3.77}$$

$$E_a = V_a + (Z_y + Z_n) I_a + Z_n I_b + Z_n I_c \tag{3.78}$$

Similarly, the expression of other voltages can be found as,

$$E_b = V_b + Z_n I_a + (Z_y + Z_n) I_b + Z_n I_c \tag{3.79}$$

$$E_c = V_c + Z_n I_a + Z_n I_b + (Z_y + Z_n) I_c \tag{3.80}$$

Equations (3.78), (3.79) and (3.80) can be written in the matrix form as,

$$\begin{bmatrix} E_a \\ E_b \\ E_c \end{bmatrix} = \begin{bmatrix} V_a \\ V_b \\ V_c \end{bmatrix} + \begin{bmatrix} Z_y + Z_n & Z_n & Z_n \\ Z_n & Z_y + Z_n & Z_n \\ Z_n & Z_n & Z_y + Z_n \end{bmatrix} \begin{bmatrix} I_a \\ I_b \\ I_c \end{bmatrix} \tag{3.81}$$

More concisely, Eq. (3.81) can be written as,

$$[E]_{abc} = [V]_{abc} + [Z]_{abc} [I]_{abc} \tag{3.82}$$

where,

$$[E]_{abc} = [E_a \quad E_b \quad E_c]^T$$
$$[V]_{abc} = [V_a \quad V_b \quad V_c]^T$$
$$[I]_{abc} = [I_a \quad I_b \quad I_c]^T$$

Multiplying Eq. (3.82) by $[A]^{-1}$ yields,

$$[A]^{-1} [E]_{abc} = [A]^{-1} [V]_{abc} + [A]^{-1} [Z]_{abc} [I]_{abc} \tag{3.83}$$

Let us consider the generator voltage,

$$E_a = E \tag{3.84}$$

According to the a operator, the following relationships can be affirmed,

$$E_b = a^2 E \tag{3.85}$$

$$E_c = aE \tag{3.86}$$

The left hand side of Eq. (3.83) can be modified as,

$$[A]^{-1}[E]_{abc} = \frac{1}{3}\begin{bmatrix} 1 & 1 & 1 \\ 1 & a & a^2 \\ 1 & a^2 & a \end{bmatrix}\begin{bmatrix} E \\ a^2 E \\ aE \end{bmatrix} \qquad (3.87)$$

$$[A]^{-1}[E]_{abc} = \frac{E}{3}\begin{bmatrix} 1+a+a^2 \\ 1+a^3+a^3 \\ 1+a^4+a^2 \end{bmatrix} = \frac{E}{3}\begin{bmatrix} 0 \\ 3 \\ 0 \end{bmatrix} = \begin{bmatrix} 0 \\ E \\ 0 \end{bmatrix} \qquad (3.88)$$

The first part of the right hand side of Eq. (3.83) becomes,

$$[A]^{-1}[V]_{abc} = [V]_{012} \qquad (3.89)$$

The second part of the right hand side of Eq. (3.83) can be written as,

$$[A]^{-1}[Z]_{abc}[I]_{abc} = [A]^{-1}[Z]_{abc}[A][I]_{012} \qquad (3.90)$$

However,

$$[A]^{-1}[Z]_{abc}[A] = [Z]_{012} \qquad (3.91)$$

Substituting Eqs. (3.91) into Eq. (3.90) yields,

$$[A]^{-1}[Z]_{abc}[I]_{abc} = [Z]_{012}[I]_{012} \qquad (3.92)$$

where,

$$[Z]_{012} = \begin{bmatrix} 3Z_n+Z_y & 0 & 0 \\ 0 & Z_y & 0 \\ 0 & 0 & Z_y \end{bmatrix} = \begin{bmatrix} Z_0 & 0 & 0 \\ 0 & Z_1 & 0 \\ 0 & 0 & Z_2 \end{bmatrix} \qquad (3.93)$$

In Eq. (3.93),

$Z_0 = Z_y + 3Z_n$ is the zero sequence impedance,
$Z_1 = Z_y$ is the positive sequence impedance,
$Z_2 = Z_y$ is the negative sequence impedance.

Substituting Eqs. (3.88), (3.89), (3.92) into Eq. (3.83) yields,

$$\begin{bmatrix} 0 \\ E \\ 0 \end{bmatrix} = \begin{bmatrix} V_0 \\ V_1 \\ V_2 \end{bmatrix} + \begin{bmatrix} Z_0 & 0 & 0 \\ 0 & Z_1 & 0 \\ 0 & 0 & Z_2 \end{bmatrix}\begin{bmatrix} I_0 \\ I_1 \\ I_2 \end{bmatrix} \qquad (3.94)$$

Equation (3.94) can be modified as,

$$\begin{bmatrix} V_0 \\ V_1 \\ V_2 \end{bmatrix} = \begin{bmatrix} 0 \\ E \\ 0 \end{bmatrix} - \begin{bmatrix} Z_0 & 0 & 0 \\ 0 & Z_1 & 0 \\ 0 & 0 & Z_2 \end{bmatrix} \begin{bmatrix} I_0 \\ I_1 \\ I_2 \end{bmatrix} \tag{3.95}$$

The equations of the voltages are,

$$V_0 = 0 - I_0 Z_0 \tag{3.96}$$

$$V_1 = E - I_1 Z_1 \tag{3.97}$$

$$V_2 = 0 - I_2 Z_2 \tag{3.98}$$

The sequence networks are shown in Fig. 3.11.

3.11 Single Line-to-Ground Fault

A three-phase Y-connected unload generator is shown in Fig. 3.25. Initially, the neutral of the generator is grounded with a solid wire. In this case, consider that a single line-to-ground fault occurs in phase a of the unloaded generator which disturbs the balance of the power system network. For this scenario, the boundary conditions are,

$$V_a = 0 \tag{3.99}$$

$$I_b = 0 \tag{3.100}$$

$$I_c = 0 \tag{3.101}$$

Fig. 3.25 Unloaded generator

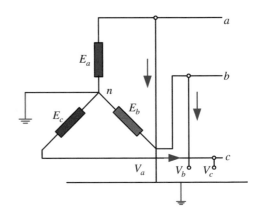

The expressions of symmetrical components of current of phase a are,

$$\begin{bmatrix} I_{a0} \\ I_{a1} \\ I_{a2} \end{bmatrix} = \frac{1}{3} \begin{bmatrix} 1 & 1 & 1 \\ 1 & a & a^2 \\ 1 & a^2 & a \end{bmatrix} \begin{bmatrix} I_a \\ I_b \\ I_c \end{bmatrix} \tag{3.102}$$

Substituting Eqs. (3.100) and (3.101) into Eq. (3.102) yields,

$$\begin{bmatrix} I_{a0} \\ I_{a1} \\ I_{a2} \end{bmatrix} = \frac{1}{3} \begin{bmatrix} 1 & 1 & 1 \\ 1 & a & a^2 \\ 1 & a^2 & a \end{bmatrix} \begin{bmatrix} I_a \\ 0 \\ 0 \end{bmatrix} \tag{3.103}$$

From Eq. (3.103), the symmetrical components of the current of phase a can be written as,

$$I_{a0} = I_{a1} = I_{a2} = \frac{I_a}{3} \tag{3.104}$$

From Eq. (3.104), it is observed that the symmetrical components of current are equal in a single line-to ground fault.

Substituting the values of symmetrical components of current in Eq. (3.95) yields,

$$\begin{bmatrix} V_{a0} \\ V_{a1} \\ V_{a2} \end{bmatrix} = \begin{bmatrix} 0 \\ E_a \\ 0 \end{bmatrix} - \begin{bmatrix} Z_0 & 0 & 0 \\ 0 & Z_1 & 0 \\ 0 & 0 & Z_2 \end{bmatrix} \begin{bmatrix} I_{a1} \\ I_{a1} \\ I_{a1} \end{bmatrix} \tag{3.105}$$

From Eq. (3.105), the symmetrical components of voltage can be written as,

$$V_{a0} = -Z_0 I_{a1} \tag{3.106}$$

$$V_{a1} = E_a - Z_1 I_{a1} \tag{3.107}$$

$$V_{a2} = -Z_2 I_{a1} \tag{3.108}$$

However, we know that the unbalanced voltage for phase a is,

$$V_a = V_{a0} + V_{a1} + V_{a2} = 0 \tag{3.109}$$

Substituting Eqs. (3.106), (3.107) and (3.108) into Eq. (3.109) yields,

$$-Z_0 I_{a1} + E_a - Z_1 I_{a1} - Z_2 I_{a1} = 0 \tag{3.110}$$

$$I_{a1} = \frac{E_a}{Z_0 + Z_1 + Z_2} \tag{3.111}$$

For a single line-to-ground fault, the positive sequence current, negative sequence current and the zero sequence current are equal and can be determined from Eq. (3.111).

From Eq. (3.104), the expression of fault current in phase a can be determined as,

$$I_a = 3I_{a1} \tag{3.112}$$

Substituting Eq. (3.111) into Eq. (3.112) yields,

$$I_a = \frac{3E_a}{Z_0 + Z_1 + Z_2} \tag{3.113}$$

The sequence network connection for the single line-to-ground fault without a fault impedance is shown in Fig. 3.26.

If a single line-to-ground fault occurs in phase a through the impedance Z_f then the expression of fault current from Eq. (3.109) can be derived as,

$$V_a = V_{a0} + V_{a1} + V_{a2} = I_a Z_f \tag{3.114}$$

Substituting Eqs. (3.106), (3.107), (3.108) and (3.112) into Eq. (3.114) yields,

$$-Z_0 I_{a1} + E_a - Z_1 I_{a1} - Z_2 I_{a1} = 3Z_f I_{a1} \tag{3.115}$$

$$I_{a1} = \frac{E_a}{Z_0 + Z_1 + Z_2 + 3Z_f} \tag{3.116}$$

Again, substituting Eq. (3.116) into Eq. (3.112), the fault current in phase a can be determined as,

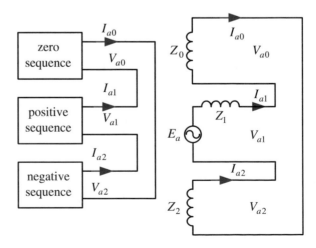

Fig. 3.26 Connection of sequence networks for single line-to-ground fault without fault impedance

Fig. 3.27 Connection of
sequence networks for single
line-to-ground fault with fault
impedance

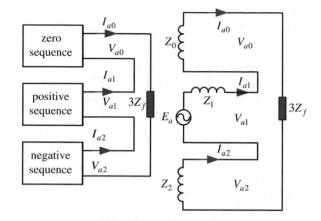

$$I_a = \frac{3E_a}{Z_0 + Z_1 + Z_2 + 3Z_f} \qquad (3.117)$$

The sequence network connection for the single line-to-ground fault with a fault
impedance is shown in Fig. 3.27.

Example 3.4 A three-phase 15 MVA, Y-connected, 11 kV synchronous generator
is solidly grounded. The positive, negative and zero sequence impedances are
$j1.5\,\Omega$, $j0.8\,\Omega$ and $j0.3\,\Omega$, respectively. Determine the fault current in phase a if the
single line-to-ground fault occurs in that phase.

Solution

The value of the generated voltage per phase is,

$$E_a = \frac{11000}{\sqrt{3}} = 6350.8 \text{ V}$$

The value of the symmetrical component of the current can be determined as,

$$I_{a1} = \frac{E_a}{Z_0 + Z_1 + Z_2} = \frac{6350.8}{j1.5 + j0.8 + j0.3} = -j2442.6 \text{ A}$$

The value of the fault current in the phase a can be calculated as,

$$I_a = 3I_{a1} = 3 \times -j2442.6 = -j7327.8 \text{ A}$$

Example 3.5 Figure 3.28 shows a Y-connected three-phase synchronous generator
whose neutral is earthed with solid wire. A single line-to-ground fault occurs in
phase a and the current in this phase is found to be 100 A. Find the positive
sequence, negative sequence and zero sequence currents for all three phases.

Fig. 3.28 Y-connected
synchronous generator

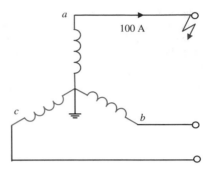

Solution

During fault, the current in different phases are,

$$I_a = 100\,\text{A}$$
$$I_b = 0\,\text{A}$$
$$I_c = 0\,\text{A}$$

The zero sequence components of the current can be determined as,

$$I_{a0} = I_{b0} = I_{c0} = \frac{1}{3}(I_a + I_b + I_c) = \frac{1}{3} \times 100 = 33.33\,\text{A}$$

The positive sequence components of the current can be determined as,

$$I_{a1} = \frac{1}{3}(I_a + aI_b + a^2I_c) = \frac{1}{3} \times 100 = 33.33\,\text{A}$$
$$I_{b1} = a^2I_{a1} = 33.33\lfloor\underline{240^\circ}\,\text{A}$$
$$I_{c1} = aI_{a1} = 33.33\lfloor\underline{120^\circ}\,\text{A}$$

The negative sequence components of the current can be determined as,

$$I_{a2} = \frac{1}{3}(I_a + a^2I_b + aI_c) = \frac{1}{3} \times 100 = 33.33\,\text{A}$$
$$I_{b2} = aI_{a2} = 33.33\lfloor\underline{120^\circ}\,\text{A}$$
$$I_{c2} = a^2I_{a2} = 33.33\lfloor\underline{240^\circ}\,\text{A}$$

Practice problem 3.4
A three-phase 20 MVA, 11 kV synchronous generator is having a sub transient
reactance of 0.20 pu and its neutral is solidly grounded. The negative and zero
sequence reactance are 0.30 and 0.15 pu, respectively. Determine the fault current if
a single line- to-ground fault occurs in phase *a*.

Fig. 3.29 A Y-connected
synchronous generator

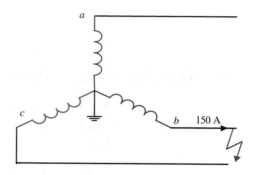

Practice problem 3.5

The neutral of a three-phase Y-connected synchronous generator is solidly grounded as shown in Fig. 3.29. A single line-to-ground fault occurs in phase b and the current in this phase is found to be 150 A. Calculate the positive sequence, negative sequence and zero sequence currents for all three phases.

3.12   Line-to-Line Fault

A three-phase synchronous generator whose line-to-line fault occurs between phase b and phase c as shown in Fig. 3.30. In this condition, the voltages at phases b and c must be the same. Whereas the currents in the phase b and phase c must be equal but in opposite direction to each other. In the line-to-line fault, the boundary conditions are,

$$I_a = 0 \qquad\qquad\qquad (3.118)$$

Fig. 3.30 Line-to-line fault
on an unloaded synchronous
generator

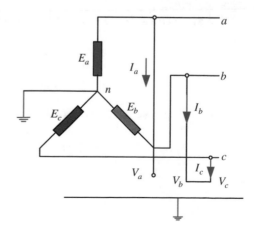

$$I_b = -I_c \tag{3.119}$$

$$V_b = V_c \tag{3.120}$$

Substituting Eqs. (3.118) and (3.119) into Eq. (3.102) yields,

$$\begin{bmatrix} I_{a0} \\ I_{a1} \\ I_{a2} \end{bmatrix} = \frac{1}{3} \begin{bmatrix} 1 & 1 & 1 \\ 1 & a & a^2 \\ 1 & a^2 & a \end{bmatrix} \begin{bmatrix} 0 \\ I_b \\ -I_b \end{bmatrix} \tag{3.121}$$

$$\begin{bmatrix} I_{a0} \\ I_{a1} \\ I_{a2} \end{bmatrix} = \frac{1}{3} \begin{bmatrix} 0 + I_b - I_b \\ 0 + (a - a^2)I_b \\ 0 + (a^2 - a)I_b \end{bmatrix} \tag{3.122}$$

From Eq. (3.122), the symmetrical components of current in phase a can be written as,

$$I_{a0} = 0 \tag{3.123}$$

$$I_{a1} = \frac{1}{3}(a - a^2)I_b \tag{3.124}$$

$$I_{a2} = \frac{1}{3}(a^2 - a)I_b \tag{3.125}$$

From Eq. (3.123), it is seen that the zero sequence component of current is zero. Hence, the value of the zero sequence voltage can be written as,

$$V_{a0} = I_{a0}Z_0 = 0 \tag{3.126}$$

From Eqs. (3.124) and (3.125), it is also seen that the positive sequence component of current is equal to the negative sequence component of current but in opposite direction.

The symmetrical components of voltage can be determined as,

$$V_b = V_{a0} + a^2 V_{a1} + a V_{a2} \tag{3.127}$$

$$V_c = V_{a0} + a V_{a1} + a^2 V_{a2} \tag{3.128}$$

Substituting Eqs. (3.127) and (3.128) into Eq. (3.120) yields,

$$V_{a0} + a^2 V_{a1} + a V_{a2} = V_{a0} + a V_{a1} + a^2 V_{a2} \tag{3.129}$$

$$(a^2 - a)V_{a1} = (a^2 - a)V_{a2} \tag{3.130}$$

$$V_{a1} = V_{a2} \tag{3.131}$$

Substituting Eqs. (3.107) and (3.108) into Eq. (3.131) yields,

$$E_a - I_{a1}Z_1 = -I_{a2}Z_2 \tag{3.132}$$

$$E_a - I_{a1}Z_1 = I_{a1}Z_2 \tag{3.133}$$

$$I_{a1} = \frac{E_a}{Z_1 + Z_2} \tag{3.134}$$

The interconnection of the sequence network without a fault impedance is shown in Fig. 3.31.

If there is a fault impedance Z_f in between lines b and c, then the following expression can be written as,

$$V_b = V_c + I_b Z_f \tag{3.135}$$

Substituting Eqs. (3.17), (3.127) and (3.128) into Eq. (3.135) yields,

$$V_{a0} + a^2 V_{a1} + a V_{a2} = V_{a0} + a V_{a1} + a^2 V_{a2} + (I_{a0} + a^2 I_{a1} + a I_{a2}) Z_f \tag{3.136}$$

Again, substituting Eqs. (3.123), (3.124) and (3.125) into Eq. (3.136) provides,

$$(a^2 - a) V_{a1} = (a^2 - a) V_{a2} + (a^2 - a) I_{a1} Z_f \tag{3.137}$$

$$V_{a1} = V_{a2} + I_{a1} Z_f \tag{3.138}$$

The interconnection of the sequence network with a fault impedance is shown in Fig. 3.32.

Fig. 3.31 Sequence network for line-to-line fault without fault impedance

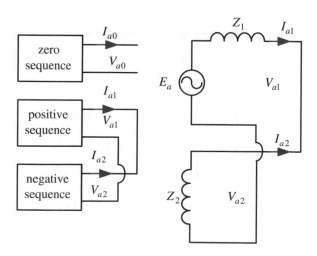

Fig. 3.32 Sequence network for line-to-line fault with a fault impedance

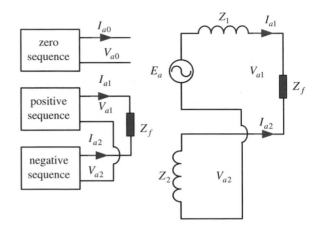

Example 3.6 The positive sequence, negative sequence and zero sequence reactance of a 15 MVA, 13 kV three-phase Y-connected synchronous generator are 0.4, 0.3 and 0.1 pu, respectively. The neutral point of the generator is solidly grounded and is not supplying current to the load. Calculate the fault current and the actual line-to-line voltages if a line-to-line fault occurs between phase b and phase c.

Solution

Let us consider that the base values are 15 MVA and 13 kV. The per unit generator voltage can be determined as,

$$E = \frac{13}{13} = 1\underline{|0°}\,\text{pu}$$

The zero sequence component of current is,

$$I_{a0} = 0$$

The values of the positive and negative sequence components of current are,

$$I_{a1} = -I_{a2} = \frac{E}{Z_1 + Z_2} = \frac{1}{j0.4 + j0.3} = -j1.42\,\text{pu}$$

The value of the fault current can be determined as,

$$I_b = I_{a0} + a^2 I_{a1} + a I_{a2} = 0 + 1\underline{|240°} \times 1.42\underline{|-90°} + 1\underline{|120°} \times 1.42\underline{|90°}$$

$$I_b = 0 + 1.42\underline{|150°} + 1.42\underline{|210°} = -2.44\,\text{pu}$$

The value of the base current can be calculated as,

$$I_{base} = \frac{15 \times 1000000}{\sqrt{3} \times 13 \times 1000} = 666.17\,\text{A}$$

The actual value of the fault current is,

$$I_b = 2.44 \times 666.17 = 1625.45\,\text{A}$$

The sequence voltages for phase a can be determined as,

$$V_{a0} = -I_{a0}Z_0 = 0\underline{|0°}\ \text{pu}$$
$$V_{a1} = E - I_{a1}Z_1 = 1\underline{|0°} + j1.42 \times j0.4 = 0.432\underline{|0°}\ \text{pu}$$
$$V_{a2} = -I_{a2}Z_2 = -j1.42 \times j0.3 = 0.436\underline{|0°}\ \text{pu}$$

The phase voltages of the generator are calculated as,

$$V_a = V_{a0} + V_{a1} + V_{a2} = 0 + 0.432 + 0.436 = 0.87\underline{|0°}\ \text{pu}$$
$$V_b = V_{b0} + a^2 V_{a1} + a V_{a2} = 0 + 0.432\underline{|240°} + 0.436\underline{|120°} = \underline{|0°}\ \text{pu}$$
$$V_c = V_{c0} + a V_{a1} + a^2 V_{a2} = 0 + 0.432\underline{|120°} + 0.436\underline{|240°} = \underline{|0°}\ \text{pu}$$

The line-to-line voltages of the generator are determined as,

$$V_{ab} = V_a - V_b = 0.87\underline{|0°} - \underline{|0°} = 0.87\underline{|0°}\ \text{pu}$$

$$V_{bc} = V_b - V_c = 0\underline{|0°} - 0\underline{|0°} = 0\underline{|0°}\ \text{pu}$$

$$V_{ca} = V_c - V_a = 0\underline{|0°} - 0.87\underline{|0°} = 0.87\underline{|-90°}\ \text{pu}$$

The value of the line to neutral voltages is calculated as,

$$V_{Ln} = \frac{13}{\sqrt{3}} = 7.51\,\text{kV}$$

The actual line-to-line voltages are calculated as,

$$V_{ab} = 7.51 \times 0.87\underline{|0°} = 6.53\underline{|0°}\ \text{kV}$$
$$V_{bc} = 7.51 \times 0\underline{|0°} = 0\underline{|0°}\ \text{kV}$$
$$V_{ca} = 7.51 \times 0.87\underline{|-90°} = 6.53\underline{|-90°}\ \text{kV}$$

Practice problem 3.6

The positive sequence, negative sequence and zero sequence components of reactance of a 20 MVA, 13.8 kV synchronous generator are 0.3, 0.2 and 0.1 pu, respectively. The neutral point of the generator is solidly grounded and the generator is not supplying current to load. Find the line-to-line voltages of the generator if fault occurs between phases b and c.

3.13 Double Line-to-Ground Fault

The connection diagram of an unloaded synchronous generator is shown in Fig. 3.33. Consider that a double line-to-ground fault occurs between phases b and c. The voltages at phase b and phase c should be equal to zero and the current in the phase a is equal to zero. In this case, the boundary conditions are,

$$I_a = 0 \tag{3.139}$$

$$V_b = V_c = 0 \tag{3.140}$$

According to Eq. (3.24), the symmetrical components of voltage for phase a can be expressed as,

$$\begin{bmatrix} V_{a0} \\ V_{a1} \\ V_{a2} \end{bmatrix} = \frac{1}{3} \begin{bmatrix} 1 & 1 & 1 \\ 1 & a & a^2 \\ 1 & a^2 & a \end{bmatrix} \begin{bmatrix} V_a \\ V_b \\ V_c \end{bmatrix} \tag{3.141}$$

Substituting Eq. (3.140) into Eq. (3.141) yields,

$$\begin{bmatrix} V_{a0} \\ V_{a1} \\ V_{a2} \end{bmatrix} = \frac{1}{3} \begin{bmatrix} 1 & 1 & 1 \\ 1 & a & a^2 \\ 1 & a^2 & a \end{bmatrix} \begin{bmatrix} V_a \\ 0 \\ 0 \end{bmatrix} \tag{3.142}$$

$$V_{a0} = V_{a1} = V_{a2} = \frac{V_a}{3} \tag{3.143}$$

Initially,

$$V_{a0} = V_{a1} \tag{3.144}$$

Fig. 3.33 Double line-to-line fault on an unloaded generator

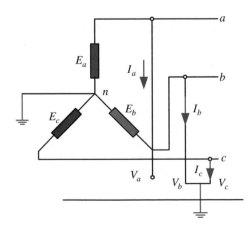

Substituting Eqs. (3.96) and (3.97) into Eq. (3.144) yields,

$$-I_{a0}Z_0 = E_a - I_{a1}Z_1 \tag{3.145}$$

$$I_{a0} = -\frac{E_a - I_{a1}Z_1}{Z_0} \tag{3.146}$$

Finally,

$$V_{a2} = V_{a1} \tag{3.147}$$

Substituting Eqs. (3.97) and (3.98) into Eq. (3.147) yields,

$$E_a - I_{a1}Z_1 = -I_{a2}Z_2 \tag{3.148}$$

$$I_{a2} = -\frac{E_a - I_{a1}Z_1}{Z_2} \tag{3.149}$$

The unsymmetrical current for phase a is,

$$I_a = I_{a0} + I_{a1} + I_{a2} \tag{3.150}$$

Substituting Eqs. (3.139), (3.146) and (3.149) into Eq. (3.150) yields,

$$0 = -\frac{E_a - Z_1 I_{a1}}{Z_0} + I_{a1} - \frac{E_a - Z_1 I_{a1}}{Z_2} \tag{3.151}$$

$$I_{a1}\left(1 + \frac{Z_1}{Z_0} + \frac{Z_1}{Z_2}\right) = E_a\left(\frac{1}{Z_0} + \frac{1}{Z_2}\right) \tag{3.152}$$

$$I_{a1} = \frac{E_a\left(\frac{Z_2 + Z_0}{Z_0 Z_2}\right)}{\left(1 + \frac{Z_1 Z_2 + Z_1 Z_0}{Z_0 Z_2}\right)} \tag{3.153}$$

$$I_{a1} = \frac{E_a\left(\frac{Z_2 + Z_0}{Z_0 Z_2}\right)}{\left(\frac{Z_0 Z_2 + Z_1 Z_2 + Z_1 Z_0}{Z_0 Z_2}\right)} \tag{3.154}$$

$$I_{a1} = \frac{E_a(Z_2 + Z_0)}{Z_0 Z_2 + Z_1 Z_2 + Z_1 Z_0} \tag{3.155}$$

$$I_{a1} = \frac{E_a(Z_2 + Z_0)}{Z_0 Z_2 + Z_1(Z_2 + Z_0)} \tag{3.156}$$

Fig. 3.34 Sequence network for double line-to-line fault without a fault impedance

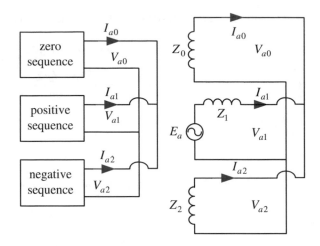

$$I_{a1} = \frac{E_a}{Z_1 + \frac{Z_0 Z_2}{Z_2 + Z_0}} \tag{3.157}$$

From Eq. (3.157), it is observed that the zero sequence impedance and negative sequence impedance are connected in parallel, then it is connected in series with the positive sequence impedance as shown in Fig. 3.34. By applying current divider rule to the circuit shown in Fig. 3.34 the negative sequence and positive sequence currents can be found as,

$$I_{a2} = -I_{a1} \frac{Z_0}{Z_2 + Z_0} \tag{3.158}$$

$$I_{a0} = -I_{a1} \frac{Z_2}{Z_2 + Z_0} \tag{3.159}$$

Again, consider that the double line to ground fault occurs between phases b and c through the ground impedance Z_f as shown in Fig. 3.35. The voltage between the fault terminal and the ground is,

$$V_b = V_c = (I_b + I_c)Z_f \tag{3.160}$$

According to Eqs. (3.17) and (3.18), the following equations can be derived,

$$V_b = V_{a0} + a^2 V_{a1} + a V_{a2} \tag{3.161}$$

$$V_c = V_{a0} + a V_{a1} + a^2 V_{a2} \tag{3.162}$$

Fig. 3.35 Double line-to-line fault on an unloaded generator with fault impedance

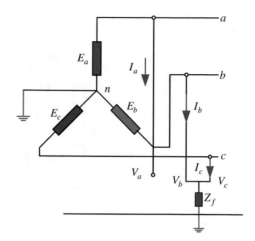

Substituting Eqs. (3.17), (3.18) and (3.161) into Eq. (3.160) yields,

$$V_{a0} + a^2 V_{a1} + a V_{a2} = (I_{a0} + a^2 I_{a1} + a I_{a2} + I_{a0} + a I_{a1} + a^2 I_{a2}) Z_f \qquad (3.163)$$

$$V_{a0} + a^2 V_{a1} + a V_{a2} = 2 I_{a0} Z_f + (a^2 + a) I_{a1} Z_f + (a + a^2) I_{a2} Z_f \qquad (3.164)$$

$$V_{a0} + a^2 V_{a1} + a V_{a2} = 2 I_{a0} Z_f - I_{a1} Z_f - I_{a2} Z_f \qquad (3.165)$$

Again, substituting Eq. (3.143) into Eq. (3.165) yields,

$$V_{a0} + (a^2 + a) V_{a1} = 2 I_{a0} Z_f - (I_{a1} + I_{a2}) Z_f \qquad (3.166)$$

$$V_{a0} - V_{a1} = 2 I_{a0} Z_f - (I_{a1} + I_{a2}) Z_f \qquad (3.167)$$

Substituting Eq. (3.139) into Eq. (3.150) yields,

$$0 = I_{a0} + I_{a1} + I_{a2} \qquad (3.168)$$

$$I_{a1} + I_{a2} = -I_{a0} \qquad (3.169)$$

Again, substituting Eq. (3.169) into Eq. (3.167) yields,

$$V_{a0} - V_{a1} = 3 I_{a0} Z_f \qquad (3.170)$$

The sequence network with the fault impedance is shown in Fig. 3.36.

Example 3.7 The positive sequence, negative sequence and zero sequence reactance of a 20 MVA, 11 kV three-phase Y-connected synchronous generator are 0.24, 0.19 and 0.18 pu, respectively. The generator's neutral point is solidly grounded. A double line-to-line fault occurs between phases *b* and *c*. Find the

Fig. 3.36 Sequence network
for double line-to-line fault
with a fault impedance

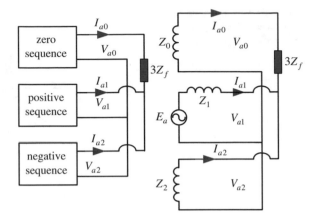

(i) currents in each phase during the sub transient period immediately after the fault occurs, and (ii) line-to-line voltages.

Solution

Consider that the base values are 20 MVA and 11 kV. Then, the per unit generator voltage is,

$$E = \frac{11}{11} = 1\underline{|0^\circ} \text{ pu}$$

The value of the positive sequence component of the current is determined as,

$$I_{a1} = \frac{E}{\left(Z_1 + \frac{Z_0 Z_2}{Z_0 + Z_2}\right)} = \frac{1}{j0.24 + \frac{j0.18 \times j0.19}{j(0.18 + 0.19)}} = -j3.01 \text{ pu}$$

The value of the negative sequence component of the current is calculated as,

$$I_{a2} = -I_{a1} \frac{Z_0}{Z_2 + Z_0} = j3.01 \frac{j0.18}{j(0.19 + 0.18)} = j1.46 \text{ pu}$$

The value of the zero sequence component of the current can be determined as,

$$I_{a0} = -I_{a1} \frac{Z_2}{Z_2 + Z_0} = j3.01 \frac{j0.19}{j(0.19 + 0.18)} = j1.55 \text{ pu}$$

(i) The value of the base current is calculated as,

$$I_b = \frac{20 \times 1000}{\sqrt{3} \times 11} = 1049.73\,\text{A}$$

The per unit values of the phase currents can be calculated as,

$$I_a = I_{a0} + I_{a1} + I_{a2} = j1.55 - j3.01 + j1.46 = 0\,\text{pu}$$
$$I_b = I_{b0} + I_{b1} + I_{b2} = I_{a0} + a^2 I_{a1} + a I_{a2}$$
$$I_b = j1.55 - 3.01\underline{|90° + 240°} + 1.46\underline{|90° + 120°} = 4.52\underline{|149.01°}\,\text{pu}$$
$$I_c = I_{c0} + I_{c1} + I_{c2} = I_{a0} + a I_{a1} + a^2 I_{a2}$$
$$I_c = j1.55 - 3.01\underline{|90° + 120°} + 1.46\underline{|90° + 240°} = 4.52\underline{|30.99°}\,\text{pu}$$

The value of the actual fault current during the sub-transient period is,

$$I_b = I_c = 4.52 \times 1049.73 = 4744.78\,\text{A}$$

(ii) The value of the sequence voltages of phase a can be determined as,

$$V_{a1} = E - I_{a1}Z_1 = 1 + j3.01 \times j0.24 = 0.28\,\text{pu}$$

In the double line-to-ground fault, the symmetrical components of voltage are the same and it can be written as,

$$V_{a0} = V_{a1} = V_{a2} = 0.28\,\text{pu}$$

The values of the phase voltages of the generator can be determined as,

$$V_a = V_{a0} + V_{a1} + V_{a2} = 0.28 + 0.28 + 0.28 = 0.84\underline{|0°}\,\text{pu}$$
$$V_b = V_{b0} + V_{b1} + V_{b2} = V_{a0} + a^2 V_{a1} + a V_{a2}$$
$$V_b = 0.28 + 0.28\underline{|240°} + 0.28\underline{|120°} = 0\underline{|0°}\,\text{pu}$$
$$V_c = V_{c0} + V_{c1} + V_{c2} = V_{a0} + a V_{a1} + a^2 V_{a2}$$
$$V_c = 0.28 + 0.28\underline{|120°} + 0.28\underline{|240°} = 0\underline{|0°}\,\text{pu}$$

The per unit line-to-line voltages are determined as,

$$V_{ab} = V_a - V_b = 0.84\underline{|0°}\,\text{pu}$$
$$V_{bc} = V_b - V_c = 0\underline{|0°}\,\text{pu}$$
$$V_{ca} = V_c - V_a = -0.84\underline{|0°} = 0.84\underline{|180°}\,\text{pu}$$

140   3   Symmetrical and Unsymmetrical Faults

The value of the phase voltage is determined as,

$$V_p = \frac{11}{\sqrt{3}} = 6.35\,\text{kV}$$

The actual line-to-line voltages of the generator can be determined as,

$$V_{ab} = 0.84 \times 6.35 = 5.33\,\text{kV}$$
$$V_{bc} = 0 \times 6.35 = 0\,\text{kV}$$
$$V_{ca} = 0.84 \times 6.35 = 5.33\,\text{kV}$$

Practice problem 3.7
The positive sequence, negative sequence and zero sequence reactance of a 50 MVA, 13.8 kV, three-phase Y-connected synchronous generator are 0.22, 0.17 and 0.15 pu, respectively. The neutral of the generator is solidly grounded. A double line-to-line fault occurs between phases b and c. Calculate the currents in each phase during the sub-transient period immediately after the fault occurs.

Example 3.8 Figure 3.37 shows a single-line diagram of a three-phase power system. The ratings of the equipment are as follows.

Generator G_1: 100 MVA, 11 kV, $X_1 = X_2 = 0.25$ pu, $X_0 = 0.05$ pu
Generator G_2: 80 MVA, 11 kV, $X_1 = X_2 = 0.15$ pu, $X_0 = 0.07$ pu
Transformer T_1: 100 MVA, 11/66 kV, $X_1 = X_2 = X_0 = 0.09$ pu
Transformer T_2: 80 MVA, 11/66 kV, $X_1 = X_2 = X_0 = 0.09$ pu
Line: $X_1 = X_2 = 15\,\Omega$, $X_0 = 30\,\Omega$

Answer the following questions by considering the power system is initially unloaded.

  (i) Draw the positive sequence, negative sequence and zero sequence networks. Also, find the equivalent sequence impedances.
 (ii) A single line-to-ground fault occurs in line a at bus 3. Find the sub-transient fault current.
(iii) A line-to-line fault occurs in lines b and c at bus 3. Calculate the sub-transient fault current.
(iv) A double-line-to ground fault occurs in lines b and c at bus 3. Determine the sub-transient fault current.

Fig. 3.37 Single-line diagram of a simple power system

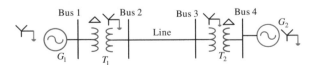

Solution

Consider that the base values are 100 MVA and 11 kV for low voltage side and 66 kV for high voltage side of the transformer. Based on the common base, the new reactance for the generators and transformers can be determined in the following ways.

The new reactance for the generator G_1 are,

$$X_1 = X_2 = \frac{100}{100} \times 0.25 = 0.25 \, \text{pu}$$

$$X_0 = \frac{100}{100} \times 0.05 = 0.05 \, \text{pu}$$

The new reactance for the generator G_2 are,

$$X_1 = X_2 = \frac{100}{80} \times 0.15 = 0.1875 \, \text{pu}$$

$$X_0 = \frac{100}{80} \times 0.07 = 0.0875 \, \text{pu}$$

The new reactance for the transformer T_1 are,

$$X_1 = X_2 = X_0 = \frac{100}{100} \times 0.09 = 0.09 \, \text{pu}$$

The new reactance for the transformer T_2 are,

$$X_1 = X_2 = X_0 = \frac{100}{80} \times 0.09 = 0.1125 \, \text{pu}$$

The base voltage for the line can be calculated as,

$$V_b = \frac{11}{a} = 11 \times \frac{66}{11} = 66 \, \text{kV}$$

The value of the base impedance can be determined as,

$$Z_b = \frac{V_b^2}{S_b} = \frac{66^2}{100} = 43.56 \, \Omega$$

The values of the new line reactance are determined as,

$$X_1 = X_2 = \frac{15}{43.56} = 0.3443 \, \text{pu}$$

$$X_0 = \frac{30}{43.56} = 0.6887 \, \text{pu}$$

(i) Figure 3.38 shows the sequence networks. The equivalent positive sequence and negative sequence impedances can be determined as,

$$Z_1 = Z_2 = \frac{j(0.25 + 0.09 + 0.3443) \times j(0.1125 + 0.1875)}{j(0.6843 + 0.3)} = j0.2085 \, \text{pu}$$

The value of the equivalent zero sequence impedance can be calculated as,

$$Z_0 = \frac{j(0.09 + 0.6887) \times j(0.1125)}{j(0.7787 + 0.1125)} = j0.0982 \, \text{pu}$$

(ii) Figure 3.39 shows the sequence network for a single line-to-ground fault. The sequence components of the current are determined as,

$$I_{a1} = I_{a2} = I_{a0} = \frac{E}{Z_1 + Z_2 + Z_0}$$

$$I_{a1} = I_{a2} = I_{a0} = \frac{1 \underline{|0^\circ}}{j0.2085 + j0.2085 + j0.0982} = 1.94 \underline{|-90^\circ} \, \text{pu}$$

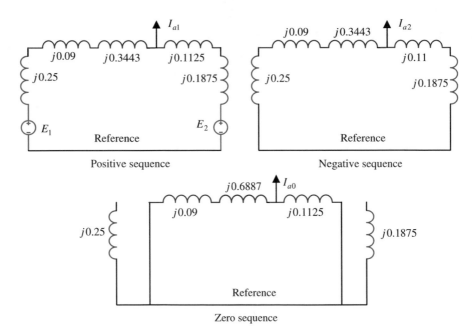

Fig. 3.38 Sequence networks

Fig. 3.39 A sequence networks

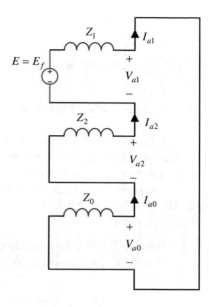

The value of the fault current in the phase a can be determined as,

$$I_a = I_{a1} + I_{a2} + I_{a0} = 1.94\underline{|-90°} + 1.94\underline{|-90°} + 1.94\underline{|-90°} = 5.82\underline{|-90°} \text{ pu}$$

The value of the base current can be calculated as,

$$I_{base} = \frac{100 \times 1000}{\sqrt{3} \times 66} = 874.77 \text{ A}$$

For a single line-to-ground fault, the actual value of the fault current can be determined as,

$$I_{f(s-l-g)} = 5.82 \times 874.77 = 5091.16 \text{ A}$$

Figure 3.40 shows the simulation result by IPSA software and the value of the fault current in the busbar 3 is found to be 5091.53 A.

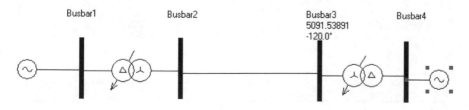

Fig. 3.40 Simulation results by IPSA software

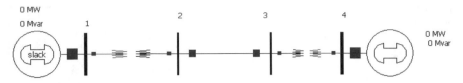

Fig. 3.41 Powerworld simulation network

The "Powerworld" software is also used to calculate the fault current as shown in Fig. 3.41. The values of the sequence reactance of the generators, transformers and transmission line are placed in the respective places. Figure 3.42 shows the simulation result for the single line-to-ground fault and its value is found to be 5.68 pu. The results is found to be 5.76 pu which is approximately the same as that of the simulation result.

(iii) Figure 3.43 shows the sequence network for the line-to-line fault. In this case, the value of the positive sequence current can be determined as,

$$I_{a1} = \frac{E}{Z_1 + Z_2} = \frac{1\underline{|0^\circ}}{j0.2085 + j0.2085} = 2.398\underline{|-90^\circ}\ \text{pu}$$

The value of the negative sequence current is determined as,

$$I_{a2} = -I_{a1} = -2.398\underline{|-90^\circ}\ \text{pu} = 2.398\underline{|90^\circ}\ \text{pu}$$

The value of the zero sequence current is,

$$I_{a0} = 0\underline{|0^\circ}\ \text{pu}$$

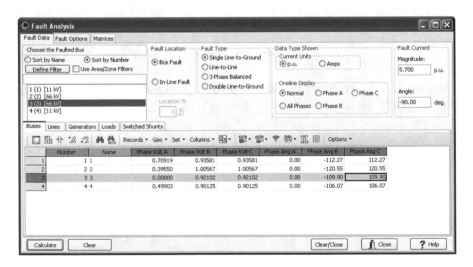

Fig. 3.42 Powerworld simulation result for SLG fault

Fig. 3.43 Sequence network
for line-to-line fault

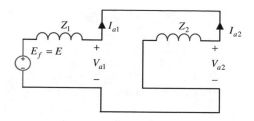

The value of the fault current in phase b is calculated as,

$$I_b = I_{b0} + I_{b1} + I_{b2} = I_{a0} + a^2 I_{a1} + a I_{a2}$$
$$I_b = 2.398\underline{|240° - 90°} + 2.398\underline{|120° + 90°} = 4.153\underline{|180°} \text{ pu}$$

The value of the fault current in phase c can be calculated as,

$$I_c = -I_b = -4.153\underline{|180°} = 4.153\underline{|180° - 180°} = 4.153\underline{|0°} \text{ pu}$$

The actual value of the fault current for line-to-line fault can be determined as,

$$I_{fl-l} = 4.153 \times 874.77 = 3632.919 \text{ A}$$

Figure 3.44 shows the IPSA simulation result for line-to-line fault and the value
of the fault current is found to be,

$$I_{f(l-l)IPSA} = 3632.33 \text{ A}$$

Figure 3.45 shows the powerworld simulation result for line-to-line fault and per
unit magnitude of the fault current is the same.

(iv) Figure 3.46 shows the sequence network for the double line-to-ground fault
and the value of the positive sequence current can be determined as,

$$I_{a1} = \frac{E}{Z_1 + \frac{Z_2 Z_0}{Z_2 + Z_0}} = \frac{1\underline{|0°}}{j0.2085 + j\frac{0.2085 \times 0.0982}{0.3067}} = 3.633\underline{|-90°} \text{ pu}$$

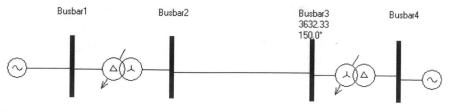

Fig. 3.44 IPSA simulation result for the line-to-line fault

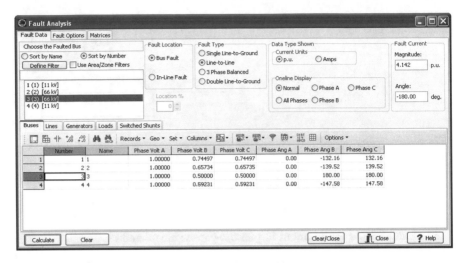

Fig. 3.45 Powerworld simulation result for the line-to-line fault

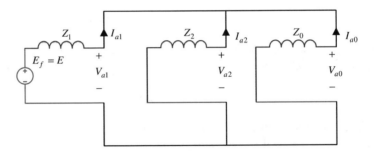

Fig. 3.46 Sequence network for the double line-to-ground fault

The value of the negative sequence current can be determined as,

$$I_{a2} = -I_{a1}\frac{Z_0}{Z_0 + Z_2} = -3.633\underline{|-90^\circ} \times \frac{j0.0982}{j0.3067} = 1.163\underline{|90^\circ}\ \text{pu}$$

The value of the zero sequence current can be calculated as,

$$I_{a0} = -I_{a1}\frac{Z_2}{Z_0 + Z_2} = -3.633\underline{|-90^\circ} \times \frac{j0.2085}{j0.3067} = 2.469\underline{|90^\circ}\ \text{pu}$$

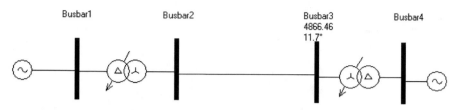

Fig. 3.47 IPSA simulation result for the double line-to-ground fault

The value of the fault current in phase b is determined as,

$$I_b = I_{b0} + I_{b1} + I_{b2} = I_{a0} + a^2 I_{a1} + a I_{a2}$$
$$I_b = 2.469\underline{|90°} + 3.633\underline{|-90° + 240°} + 1.163\underline{|90° + 120°} = 5.565\underline{|138.27°} \text{ pu}$$

The value of the fault current in phase c can be determined as,

$$I_c = I_{c0} + I_{c1} + I_{c2} = I_{a0} + a I_{a1} + a^2 I_{a2}$$
$$I_c = 2.469\underline{|90°} + 3.633\underline{|-90° + 120°} + 1.163\underline{|90° + 240°}$$
$$I_c = 5.565\underline{|41.73°} \text{ pu}$$

The actual value of the fault current can be determined as,

$$I_{f(d-l-g)} = 5.565 \times 874.77 = 4868.095 \text{ A}$$

Figure 3.47 shows the IPSA simulation result for the double line-to-ground fault and the value of the fault current is found to be,

$$I_{f(dl-g)IPSA} = 4866.46 \text{ A}$$

Exercise Problems

3.1 Calculate the quantities for (i) a^8, (ii) $a^{10} + a - 3$, and (iii) $a^{12} + 3a - 2$ by considering $a = 1\underline{|120°}$ and $a^2 = 1\underline{|240°}$.

3.2 A three-phase system is having the currents of $I_a = 10\underline{|30°}$ A, $I_b = 15\underline{|40°}$ A and $I_c = 20\underline{|35°}$ A. Calculate the symmetrical components of current in phases a and b.

3.3 A three-phase system is having the phase voltages of $V_a = 100\underline{|35°}$ V, $V_b = 200\underline{|45°}$ V and $V_c = 280\underline{|55°}$ V. Find the zero sequence, positive sequence and negative sequence components of the voltage for phase a.

Fig. 3.48 Unloaded
synchronous generator

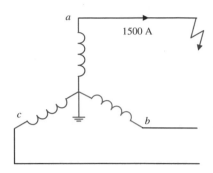

3.4 A three-phase system is having the symmetrical components of voltage of $V_{a0} = 75\underline{|45°}$ V, $V_{a1} = 155\underline{|64°}$ V and $V_{a2} = 325\underline{|85°}$ V for phase a. Calculate the phase voltages V_a, V_b and V_c.

3.5 The unbalanced currents of a three-phase system are $I_a = 50$ A, $I_b = 30 + j50$ A, $I_c = 40 + j70$ A. Calculate the zero, positive and negative sequence components of the current in phase b.

3.6 A three-phase system is having the symmetrical components of the current of $I_{a0} = 4.54 + j3.5$ A $I_{a1} = 5.34 + j1.45$ A and $I_{a2} = 1.67 - j1.85$ A for phase a. Calculate the unbalanced currents I_a, I_b and I_c if the total neutral current of this system is zero.

3.7 Figure 3.48 shows a three-phase wye-connected unloaded synchronous generator. A single line to ground fault occurs in phase a and the current in this phase is found to be 1500 A. Calculate the symmetrical components of current in phase b.

3.8 A source delivers power to a delta-connected load as shown in Fig. 3.49. The current in phase a is found to be 45 A and phase b is open circuited. Calculate the symmetrical components of the currents in all three phases.

3.9 The positive sequence, negative sequence and zero sequence reactance of a 30 MVA, 11 kV three-phase synchronous generator are measured to be 0.5, 0.4 and 0.22 pu, respectively. The generator's neutral is solidly grounded. A single line-to-ground fault occurs in phase a. Find the fault current.

3.10 A 15 MVA, 13.8 kV three-phase synchronous generator is having the positive sequence, negative sequence and zero sequence reactance of 0.3, 0.2 and 0.1 pu, respectively. The generator's neutral is solidly grounded and line-to-line fault occurs in phases b and c. Calculate the (i) fault current, (ii) sequence voltages for phase a, and (iii) phase voltages of the generator.

Fig. 3.49 Delta-connected
load

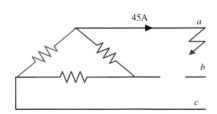

Fig. 3.50 A single-line
diagram of a power system

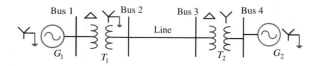

Fig. 3.51 A single-line
diagram of a power system

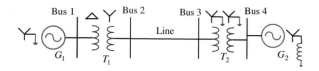

3.11 A 25 MVA, 13.8 kV three-phase Y-connected synchronous generator is
having the positive sequence, negative sequence and zero sequence reactance
of 0.34, 0.22 and 0.15 pu, respectively. The generator's neutral is solidly
grounded and the double line-to-line fault occurs between phases b and
c. Calculate the currents in each phase during the sub-transient period
immediately after the fault occurs.

3.12 Figure 3.50 shows a single-line diagram of a three-phase power system and
the ratings of the equipment are shown below.

Generators G_1, G_2: 100 MVA, 20 kV, $X_1 = X_2 = 0.20$ pu, $X_0 = 0.06$ pu
Transformers T_1, T_2: 100 MVA, 20/138 kV, $X_1 = X_2 = X_0 = 0.08$ pu
Line: 100 MVA, $X_1 = X_2 = 0.11$ pu, $X_0 = 0.55$ pu
A fault occurs at bus 4. Determine the fault currents in the faulted bus by any
simulation software (IPSA/Powerworld) for the (i) single line-to-ground,
(ii) line-to-line, and (iii) double line-to-ground faults.

3.13 A single-line diagram of a three-phase power system is shown in Fig. 3.51.
The ratings of the equipment are shown below.

Generator G_1: 100 MVA, 11 kV, $X_1 = X_2 = 0.20$ pu, $X_0 = 0.05$ pu
Generator G_2: 100 MVA, 20 kV, $X_1 = X_2 = 0.25$ pu, $X_0 = 0.03$ pu,
$X_n = 0.05$ pu
Transformer T_1: 100 MVA, 11/66 kV, $X_1 = X_2 = X_0 = 0.06$ pu
Transformer T_2: 100 MVA, 11/66 kV, $X_1 = X_2 = X_0 = 0.06$ pu
Line: 100 MVA, $X_1 = X_2 = 0.15$ pu, $X_0 = 0.65$ pu
A single line-to-ground, line-to-line, and double line-to-ground fault occur at
bus 3. Find the fault currents in each case.

3.14 Figure 3.52 shows a single-line diagram of a three-phase power system and
the ratings of the equipment are shown below.

Generator G_1: 100 MVA, 11 kV, $X_1 = X_2 = 0.20$ pu, $X_0 = 0.05$ pu
Generator G_2: 100 MVA, 20 kV, $X_1 = X_2 = 0.25$ pu, $X_0 = 0.03$ pu, $X_n = 0.05$ pu
Generator G_3: 100 MVA, 20 kV, $X_1 = X_2 = 0.30$ pu, $X_0 = 0.08$ pu

Fig. 3.52 A single-line
diagram of a power system

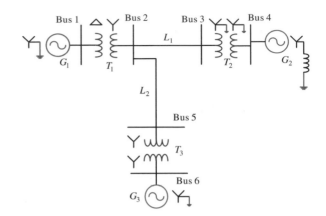

Transformer T_1: 100 MVA, 11/66 kV, Δ-Y, $X_1 = X_2 = X_0 = 0.06$ pu
Transformer T_2: 100 MVA, 20/66 kV, both sides earthed Y-Y,
$X_1 = X_2 = X_0 = 0.06$ pu
Transformer T_3: 100 MVA, 20/66 kV, Y-Y, $X_1 = X_2 = X_0 = 0.04$ pu
Line 1: 100 MVA, $X_1 = X_2 = 0.15$ pu, $X_0 = 0.65$ pu
Line 2: 100 MVA, $X_1 = X_2 = 0.10$ pu, $X_0 = 0.45$ pu
Calculate the fault currents in case of single line-to-ground, line-line and
double line-to-ground faults occurring at bus 3.

3.15 A single-line diagram of a three-phase power station is shown in Fig. 3.53.
      The ratings of the equipment are shown below.

Generator G_1: 100 MVA, 11 kV, $X_1 = X_2 = 0.24$ pu, $X_0 = 0.06$ pu
Generator G_2: 100 MVA, 20 kV, $X_1 = X_2 = 0.22$ pu, $X_0 = 0.03$ pu,
$X_n = 0.05$ pu
Generator G_3: 100 MVA, 20 kV, $X_1 = X_2 = 0.35$ pu, $X_0 = 0.07$ pu
Transformer T_1: 100 MVA, 11/66 kV, Δ-Y, $X_1 = X_2 = X_0 = 0.06$ pu

Fig. 3.53 A single-line
diagram of a power system

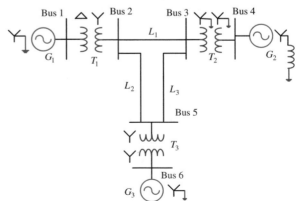

Transformer T_2: 100 MVA, 20/66 kV, both sides earthed Y-Y, $X_1 = X_2 = X_0 = 0.06$ pu

Transformer T_3: 100 MVA, 20/66 kV, Y-Y, $X_1 = X_2 = X_0 = 0.04$ pu

Line 1: 100 MVA, $X_1 = X_2 = 0.15$ pu, $X_0 = 0.65$ pu

Line 2: 100 MVA, $X_1 = X_2 = 0.10$ pu, $X_0 = 0.45$ pu

Line 3: 100 MVA, $X_1 = X_2 = 0.16$ pu, $X_0 = 0.65$ pu

Find the fault currents in each of single line-to-ground, line-line and double line-to-ground faults occurring at bus 5.

References

1. H. Sadat, *Power System Analysis*, International Edition (McGraw-Hill, New York, USA, 1999)
2. M.A. Salam, *Fundamentals of Power Systems* (Alpha Science International Ltd., Oxford, UK, 2009)
3. A.E. Guile, W. Paterson, *Electrical Power Systems*, 2nd edn. (Pergamon Press, UK, 1977)
4. T. Wildi, *Electrical Machines, Drives and Power Systems* (Pearson International Education, New Jersey, USA, 2006)
5. J. Duncan Glove, M.S. Sarma, *Power System Analysis and Design* 3rd edn. (Brooks/Cole Thomson Learning, USA, 2002)
6. S.J. Chapman, *Electric Machinery and Power System Fundamentals*, 1st edn. (McGraw-Hill Higher Education, New York, 2002)

Chapter 4
Grounding System Parameters and Expression of Ground Resistance

4.1 Introduction

In any electrical circuit network, the circuitry that provides a path between the parts of the circuit and the ground, is known as the grounding system. The grounding system is required for reliable operation of any electrical equipment including generator, transformer, power system tower, and other power system installation under fault conditions. In a grounding system, the study of ground resistance is very important in designing any electrical network for residential, commercial and industrial areas. Any electrical equipment needs to be grounded for safe operation. In this case, the enclosure of the equipment is grounded in such a way that the voltage on the equipment maintains the voltage of the ground. This is known as equipment grounding. The grounding is also an important for power generation, transmission and distribution systems. In the transmission systems, each leg of the transmission tower is grounded through a vertically inserted earth electrode. In an electrical substation, all high voltage equipment are grounded with a grounding grid system. In this chapter, objectives, symbols, expression of ground resistance with different types of electrode etc. have been discussed.

4.2 Objectives of Grounding System

The objectives of any grounding system are as follows:

Reduce insulation level of power system equipment: Power system equipment like transformer's neutral must be grounded. This can decrease the operating voltage and insulation level of the equipment.

Ensure personal safety: A good grounding system of a substation can ensure that the touch and step voltages meet the standard voltage levels.

© Springer Science+Business Media Singapore 2016
Md.A. Salam and Q.M. Rahman, *Power Systems Grounding*,
Power Systems, DOI 10.1007/978-981-10-0446-9_4

Eliminate electrostatic accidents: Static electric current may create interference of electronic devices and generate fire near any flammable object. A good grounding system can release the static current to the earth which can prevent that type of incident.

Reduce electromagnetic interference: The normal operation of any electronic device is interrupted due to electromagnetic interference. This type of interference can be reduced by the good grounding system.

Reduce cathodic protection current: The voltage is usually induced in the pipeline at the same corridor of the transmission lines which can harm the utility operators. The cathodic protection is used in the pipeline to mitigate high touch potential and reduce the current due to a fault condition. Therefore, the cathodic protection system of the pipeline must be grounded to release the current into the earth.

Detecting ground faults: In the substation, there are many leakage breakers and other fault leakage protection devices used in the low voltage circuits. A high magnitude of the ground fault current is required to bring the protection device into action, if there is an earth fault in the circuit. Therefore, to meet this condition, the neutral point on the secondary side of the step-down transformer must be grounded.

4.3   Grounding Symbols and Classification

In the grounding system, three symbols are used which are signal, chassis and earth as shown in Fig. 4.1.

The signal or equipotential ground is used for the reference voltage of an electrical circuit for simulation activities. The earth ground is used to release high magnitude of current into the earth or soil. The presence of voltage is observed in any system due to electrostatic voltage. This electrostatic voltage can be mitigated using the chassis ground. The earth and chassis grounds are normally used in the power system networks.

The grounding system basically provides a low resistance path for the fault and lightning currents in order to maintain the safe potential with respect to the zero potential. The single-phase to ground fault is the most common fault in the power system and it accounts for 98 % of all failures. The phase-to-phase and three-phase faults are responsible for 1.5 and 0.5 % of all failures respectively. The grounding system is constructed by burying electrodes into the soil. The electrodes are known

Fig. 4.1 Different symbols of grounding system

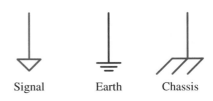

Signal   Earth   Chassis

as a 'rod' when buried vertically and 'conductor' when buried horizontally in the soil. The grounding with only rod is used in the transmission tower and distribution electric pole. The grounding grid is formed either the combination of the conductor and the rod or only the conductor. The grid is normally used in the substation and power station grounding system. Initially, grounding system is divided into grounded and ungrounded systems. The grounded system is again sub-divided into neutral grounded and non-neutral grounded systems. The neutral grounded is again divided as solidly grounding, resistance grounding, reactance grounding, voltage transformer grounding and Zig-Zag transformer grounding. The resistance grounding is sub-divided as high resistance grounding, low resistance grounding and hybrid high resistance grounding.

4.4   Ungrounded Systems

Any grounded or ungrounded systems mainly depend on the customer demands. The ungrounded electrical systems are used where the customer and the design engineer do not want the overcurrent protection device acting on the ground fault line. However, there are some ungrounded elements such as building steel or iron and, water pipeline, which are intentionally grounded. In an ungrounded system, there is no internal connection between any line (including the neutral) and ground. The ungrounded system is basically grounded through distributed capacitance. There are some advantages and disadvantages of ungrounded systems. The ground fault current in this system is very low (5 A or less) and provides more reliability during fault conditions. The voltage between the healthy lines and ground is very high. The effect of harmonics in the ungrounded system will die out itself within the system. The outside source interferences are usually neglected as there is no connection with the soil. The ungrounded system is very poor to protect the electrical appliances due to transient voltages. Sometimes, these transient voltages propagate or elongate to the nearby equipment which can destroy the insulation of those equipment. In this system, it is very difficult to locate the line to ground fault. An ungrounded wye-connection is shown in Fig. 4.2, where a ground fault occurs in line R. Each line has distributed earth capacitance. The capacitance of line R will discharge through the faulty path to the capacitance of lines Y and B. The charge and discharge of capacitance of line R will continue due to the healthy lines Y and B. The currents in the lines Y and B can be written as,

$$i_{cY} = \frac{v_{YG}}{1/j\omega C_Y} = j\omega C_Y v_{YG} \tag{4.1}$$

$$i_{cB} = \frac{v_{BG}}{1/j\omega C_B} = j\omega C_B v_{BG} \tag{4.2}$$

Fig. 4.2 Wye-connection
with faulted line R

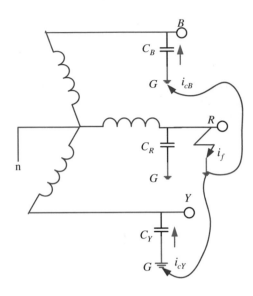

But, the distributed capacitance of each line is the same and it can be expressed as,

$$C_R = C_Y = C_B = C \tag{4.3}$$

The expression of fault current is,

$$i_f = i_{cY} + i_{cB} \tag{4.4}$$

Substituting Eqs. (4.1)–(4.3) into Eq. (4.4) provides,

$$i_f = j\omega C(v_{YG} + v_{BG}) \tag{4.5}$$

The vector diagram of the wye-connection is shown in Fig. 4.3. The current i_{cB} leads the voltage v_{BG} by 90° and The current i_{cY} leads the voltage v_{YG} by 90°. From Fig. 4.3, the following voltage expression can be written as,

$$v_{BG} = v_{Bn} + v_n \tag{4.6}$$

$$v_{YG} = v_{Yn} + v_n \tag{4.7}$$

Substituting Eqs. (4.6) and (4.7) into Eq. (4.5) yields,

$$i_f = i_R = j\omega C(v_{Yn} + v_n + v_{Bn} + v_n) \tag{4.8}$$

$$i_f = j\omega C(v_{Yn} + v_{Bn} + 2v_n) \tag{4.9}$$

Fig. 4.3 a Vector diagram of
faulted wye-connection
system. **b** Vector diagram of
faulted wye-connection
system

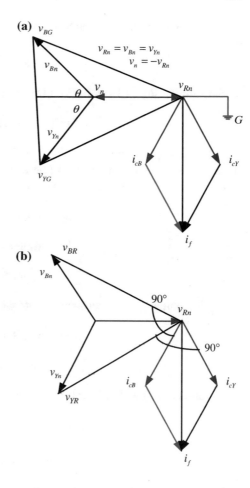

Resolve the voltage v_{Bn} into $v_{Bn} \cos \theta$ in the v_n direction and $v_{Bn} \sin \theta$ is perpendicular on the v_n. This can be expressed as,

$$v_{Bn} = v_{Bn} \cos \theta + v_{Bn} \sin \theta \qquad (4.10)$$

Again, resolving the voltage v_{Yn} into $v_{Yn} \cos \theta$ in the v_n direction and $-v_{Yn} \sin \theta$
in the perpendicular direction of the v_n, the following expression can be found,

$$v_{Yn} = v_{Yn} \cos \theta - v_{Yn} \sin \theta \qquad (4.11)$$

Adding Eqs. (4.10) and (4.11) yields,

$$v_{Yn} + v_{Bn} = v_{Bn} \cos \theta + v_{Bn} \sin \theta + v_{Yn} \cos \theta - v_{Yn} \sin \theta \qquad (4.12)$$

But, $v_{Bn} \sin \theta$ is equal and opposite to $-v_{Yn} \sin \theta$, then Eq. (4.12) can be modified as,

$$v_{Yn} + v_{Bn} = v_{Bn} \cos \theta + v_{Yn} \cos \theta = v_n \tag{4.13}$$

Substituting Eq. (4.13) into Eq. (4.9) yields,

$$i_f = j\omega C(v_n + 2v_n) = j3\omega C v_n \tag{4.14}$$

Alternative approach:
If the line to ground fault occurs at line R, then the line to ground voltage becomes zero i.e.,$v_{RG} = 0$. Applying KVL between R-G-Y in Fig. 4.2 yields,

$$v_{RG} - v_{YG} + v_{RY} = 0 \tag{4.15}$$

$$0 - v_{YG} + v_{RY} = 0 \tag{4.16}$$

$$v_{YG} = v_{RY} \tag{4.17}$$

Again, applying KVL between B-G-R yields,

$$v_{RG} - v_{BG} + v_{BR} = 0 \tag{4.18}$$

$$v_{BG} = v_{BR} \tag{4.19}$$

Equations (4.1) and (4.2) can be modified as,

$$i_{cY} = j\omega C v_{YR} = j\omega C \sqrt{3} v_{ph} \tag{4.20}$$

$$i_{cB} = j\omega C v_{BR} = j\omega C \sqrt{3} v_{ph} \tag{4.21}$$

From Fig. 4.3b, the expression of the fault current is,

$$i_f = 2i_{cB} \cos 30° = 2i_{cY} \cos 30° = \sqrt{3} i_{cB} = \sqrt{3} i_{cY} \tag{4.22}$$

Substituting Eq. (4.21) into Eq. (4.22) yields,

$$i_f = \sqrt{3} \times j\omega C \sqrt{3} v_{ph} = 3j\omega C v_{ph} \tag{4.23}$$

A delta-connection with the ungrounded system is shown in Fig. 4.4. This system is basically grounded through an earth capacitance. Applying KVL between R-G-Y yields,

$$v_{RG} - v_{YG} - v_{RY} = 0 \tag{4.24}$$

Fig. 4.4 Ungrounded
delta-connection system

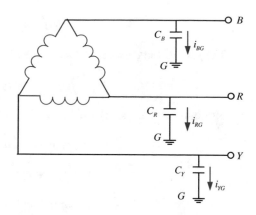

$$v_{RY} = v_{RG} - v_{YG} = 1\underline{|0°} \qquad (4.25)$$

$$v_{RG} = v_{YG} + 1\underline{|0°} \qquad (4.26)$$

Again, applying KVL between *Y-G-B* yields,

$$v_{YG} - v_{BG} - v_{YB} = 0 \qquad (4.27)$$

$$v_{YB} = v_{YG} - v_{BG} = 1\underline{|-120°} \qquad (4.28)$$

$$v_{BG} = v_{YG} - 1\underline{|-120°} \qquad (4.29)$$

Applying KVL between *B-G-R* yields,

$$v_{BG} - v_{RG} - v_{BR} = 0 \qquad (4.30)$$

$$v_{BR} = v_{BG} - v_{RG} = 1\underline{|120°} = 1\underline{|-240°} \qquad (4.31)$$

The earth capacitance will be equal for equal length and transposed of the conductor. This can be written as,

$$C_R = C_Y = C_B = C \qquad (4.32)$$

In this condition, the expressions of the current are,

$$i_{RG} = \frac{v_{RG}}{\frac{1}{j\omega C}} = j\omega C v_{RG} \qquad (4.33)$$

$$i_{YG} = j\omega C v_{YG} \qquad (4.34)$$

$$i_{BG} = j\omega C v_{BG} \tag{4.35}$$

The sum of the currents at node G can be expressed as,

$$i_{RG} + i_{YG} + i_{BG} = 0 \tag{4.36}$$

Substituting Eqs. (4.33)–(4.35) into Eq. (4.36) yields,

$$j\omega C v_{RG} + j\omega C v_{YG} + j\omega C v_{BG} = 0 \tag{4.37}$$

$$j\omega C (v_{RG} + v_{YG} + v_{BG}) = 0 \tag{4.38}$$

$$v_{RG} + v_{YG} + v_{BG} = 0 \tag{4.39}$$

Substituting Eqs. (4.26) and (4.29) into Eq. (4.39) yields,

$$v_{YG} + 1 + v_{YG} + v_{YG} - 1\underline{|-120^\circ} = 0 \tag{4.40}$$

$$v_{YG} = \frac{1}{3}(1\underline{|-120^\circ} - 1) \tag{4.41}$$

$$v_{YG} = \frac{1}{\sqrt{3}}\underline{|-150^\circ} \tag{4.42}$$

Substitute Eq. (4.42) into Eq. (4.26) provides,

$$v_{RG} = \frac{1}{\sqrt{3}}\underline{|-150^\circ} + 1\underline{|0^\circ} \tag{4.43}$$

$$v_{RG} = \frac{1}{\sqrt{3}}\underline{|-30^\circ} \tag{4.44}$$

Again, substituting Eq. (4.42) into Eq. (4.29) yields,

$$v_{BG} = \frac{1}{\sqrt{3}}\underline{|-150^\circ} - 1\underline{|-120^\circ} \tag{4.45}$$

$$v_{BG} = \frac{1}{\sqrt{3}}\underline{|90^\circ} \tag{4.46}$$

Again, consider that the line to ground fault occurs in line R as can be seen in Fig. 4.5. The impedances due to line to earth are considered to be small and the impedance due to the fault is considered to be zero as shown in Fig. 4.6. The current in the ground to line B can be written as,

Fig. 4.5 Ungrounded delta connection with fault at line R

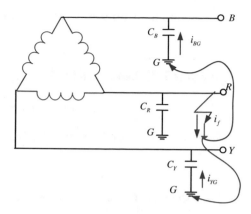

Fig. 4.6 Ungrounded delta-connection with equivalent circuit when fault at line R

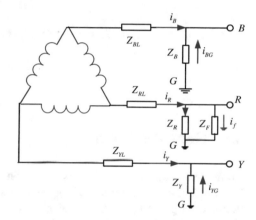

$$i_{BG} = -i_B = \frac{v_{BG}}{Z_B} \tag{4.47}$$

The current in the ground to the line Y is,

$$i_{YG} = -i_Y = \frac{v_{YG}}{Z_Y} \tag{4.48}$$

The voltage of line R to the ground is,

$$v_{RG} = 0 \tag{4.49}$$

Applying KVL between the Y-G-R yields,

$$v_{YG} - v_{RG} - v_{RY} = 0 \tag{4.50}$$

Substitute Eq. (4.49) into Eq. (4.50) provides,

$$v_{YG} = v_{YR} \qquad (4.51)$$

Applying KVL between the B-G-R yields,

$$v_{BG} - v_{RG} - v_{BR} = 0 \qquad (4.52)$$

Substituting Eq. (4.49) into Eq. (4.52) yields,

$$v_{BG} = v_{BR} \qquad (4.53)$$

The expression of the fault current can be written as,

$$i_f = i_{YG} + i_{BG} \qquad (4.54)$$

Substituting Eqs. (4.47) and (4.48) into Eq. (4.54) yields,

$$i_f = \frac{v_{YG}}{Z_Y} + \frac{v_{BG}}{Z_B} \qquad (4.55)$$

4.5 Grounded Systems

Continuous power supply is very important in the industry and commercial utilization. There are two types of a grounded system, namely, neutral grounded and non-neutral grounded. The neutral grounded system is used in the generator and transformer. The non-neutral grounded system is used for electric tower, switchgear of the substation building and utility grid substation. There are some advantages of the grounded systems. These are,

- capability of reduction in over voltages due to transient,
- greater safety for the operators,
- excellent lightning protection,
- simplified ground fault location,
- reduction in frequency of faults,
- highest system and equipment fault protection, and
- reduction in maintenance cost and time.

4.5.1 Solidly Grounded System

The neutral points of high voltage equipment like transformer and generator are solidly grounded to reduce the insulation voltage level. This grounding system is

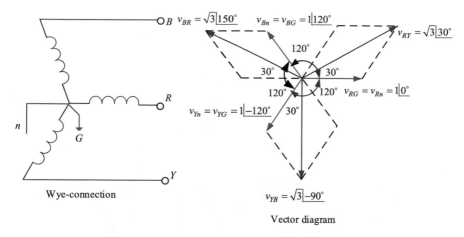

Fig. 4.7 Wye connection and vector diagram

known as working grounding. The solidly grounded system is the most widely used grounding mechanism employed in the industrial and commercial power system, where ground conductors are connected to the earth without any intermediate impedance. This grounding system has high values of current ranging from 10 to 20 kA. This current flows through the building rod, water pipe and ground wire which might damage the electrical appliances. The solidly grounded system can reduce the line to ground fault transients and locate faults easily. Some disadvantages of this grounding system are severe arc flash hazards, high fault current, observe potential on the equipment during a fault.

The wye and delta connections are usually used to step-up and step-down the input voltage into the required voltage level. The wye-connection with a solidly grounded system is shown in Fig. 4.7. In this system, the magnitudes of the phase voltages are the same. It can be written as,

$$v_{Rn} = v_{Yn} = v_{Bn} = 1\underline{|0°} \tag{4.56}$$

The expressions of the line to line voltages are,

$$v_{RY} = \sqrt{3}v_{Rn}\underline{|30°} \tag{4.57}$$

$$v_{YB} = \sqrt{3}v_{Rn}\underline{|-90°} \tag{4.58}$$

$$v_{BR} = \sqrt{3}v_{Rn}\underline{|150°} \tag{4.59}$$

The fault on an intended path of a circuit, which results in a high magnitude current that flows to the ground via an unintended shortest path is known as single-line to ground fault or in brief ground fault. Statistically, 90–95 % faults

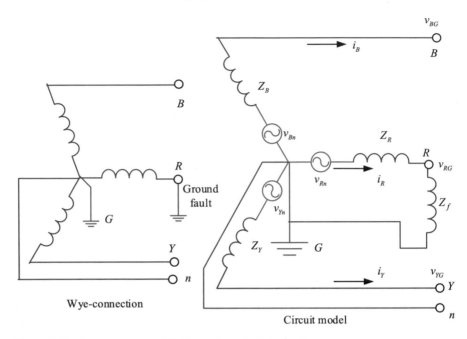

Fig. 4.8 Single-phase to ground fault and its equivalent circuit

occur due to the ground fault. The ground fault of a wye-connection and the equivalent circuit is shown in Fig. 4.8.

From the equivalent circuit in Fig. 4.8, the following equations can be written as,

$$i_R = \frac{v_{Rn}}{Z_a + Z_f} \tag{4.60}$$

$$i_Y = i_B = 0 \tag{4.61}$$

$$v_{RG} = i_R Z_f \tag{4.62}$$

Substituting Eq. (4.60) into Eq. (4.62) yields,

$$v_{RG} = \frac{v_{Rn}}{Z_a + Z_f} Z_f \tag{4.63}$$

The other phase to ground voltages are,

$$v_{YG} = v_{Yn} \tag{4.64}$$

$$v_{BG} = v_{Bn} \tag{4.65}$$

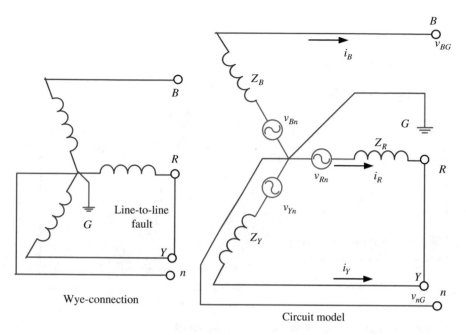

Fig. 4.9 Line-to-line fault and its equivalent circuit

The line-to-line short circuit occurs between phases R and Y of a wye-connection as shown in Fig. 4.9.

From the equivalent circuit as shown in Fig. 4.9, the equations of the current can be written as,

$$i_B = 0 \tag{4.66}$$

$$i_R = -i_Y \tag{4.67}$$

Applying KVL between the phases (lines) R and Y, and the equation can be written as,

$$-v_{Rn} + i_R Z_R - i_Y Z_Y + v_{Yn} = 0 \tag{4.68}$$

Substituting Eq. (4.67) into Eq. (4.68) yields,

$$-v_{Rn} + i_R Z_R + i_R Z_Y + v_{Yn} = 0 \tag{4.69}$$

$$i_R = \frac{v_{Rn} - v_{Yn}}{Z_R + Z_Y} \tag{4.70}$$

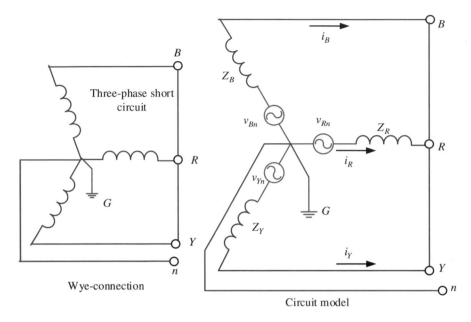

Fig. 4.10 Three-phase short circuit and its equivalent circuit

The line to line voltage between lines R and Y is,

$$V_{RY} = V_{Rn} - V_{Yn} = \sqrt{3}V_{Rn} \tag{4.71}$$

The sequence impedance of the transformer is the same, i.e.,

$$Z_R = Z_Y = Z \tag{4.72}$$

Substituting Eqs. (4.71) and (4.72) into Eq. (4.70) yields,

$$i_R = \frac{\sqrt{3}V_{Rn}}{2Z} \tag{4.73}$$

The three-phase short circuit fault is shown in Fig. 4.10. The line current can be calculated by considering any phase to neutral. For line R, the expression of the line current is,

$$i_R = \frac{V_{Rn}}{Z_R} \tag{4.74}$$

Example 4.1 A 1.5 MVA, 11/0.415 kV delta-wye three-phase transformer is connected to the power system network of a Methanol industry. The line-to ground fault occurs in the secondary of the transformer. The transformer impedance and the ground fault impedance are found to be 6 and 2 % respectively. Determine the value of the line-to-ground fault current.

Solution

The value of the line current is,

$$I_{fl} = \frac{S}{\sqrt{3}V_L} = \frac{1.5 \times 1000}{\sqrt{3} \times 0.415} = 2086.87\,\text{A} = 1\,\text{pu}$$

The pu value of the line to ground current is,

$$I_{LG} = \frac{V}{Z_T + Z_F} = \frac{1}{0.06 + 0.02} = 12.5\,\text{pu}$$

The value of the line to ground current in ampere is,

$$I_{LG} = 12.5 \times 28{,}086.87 = 351.085\,\text{kA}$$

Practice problem 4.1
A carpet industry is having a three-phase 1.0 MVA, 11/0.440 kV delta-wye transformer for its power system network. A three-phase fault occurs in the secondary of the transformer due to heavy natural wind. The transformer impedance is found to be 4 %. Determine the value of the line current.

4.5.2   Resistance Grounding

The resistance grounding system is classified as high resistance and low resistance grounding systems. The high resistance grounding system is obtained by connecting a high resistance in between the neutral point of a low voltage transformer and the ground as shown in Fig. 4.11. The high grounding system is used in the small and medium industry applications where continuous operation is required during the fault condition. The value of the resistance is chosen in such a way that it allows 1–10 A current during the ground fault.

There are some advantages of high resistance grounding. These are,

- limiting the ground fault current to a lower value,
- controlling the over transient voltages,
- maintaining continuous operation during a fault condition,
- reducing electric shock hazards,
- reducing mechanical stress in the nearby equipment, and
- reducing line voltage drop caused by occurrence and clearance of the ground fault.

Fig. 4.11 Wye-connection
with high resistance
grounding

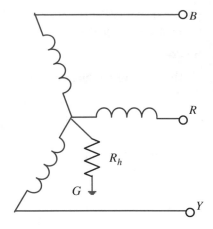

The main disadvantages are,

- high frequency signal may present during alarming period due to the fault and
- ground fault may remain present for a long time.

Consider a ground fault occurs in the line R as can be seen in Fig. 4.12. The current in the faulty phase is,

$$i_R = \frac{v_{Rn}}{Z_R + Z_f + R_h} \qquad (4.75)$$

The voltage drop across the fault impedance is,

$$v_{RG} = \frac{v_{Rn} Z_f}{Z_R + Z_f + R_h} \qquad (4.76)$$

The currents in the Y and B phases are,

$$i_Y = i_B = 0 \qquad (4.77)$$

Applying KVL in path Y-n-G provides,

$$v_{Yn} - i_R R_h - v_{YG} = 0 \qquad (4.78)$$

$$v_{YG} = v_{Yn} - i_R R_h \qquad (4.79)$$

Again, applying KVL in path B-n-G gives,

$$v_{Bn} - i_R R_h - v_{BG} = 0 \qquad (4.80)$$

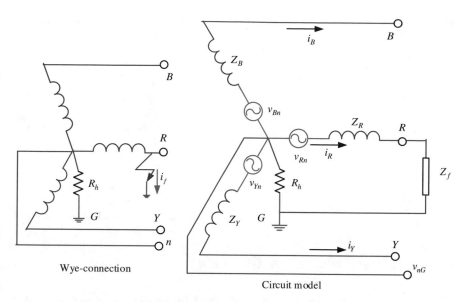

Fig. 4.12 Wye-connection with circuit model

$$v_{BG} = v_{Bn} - i_R R_h \qquad (4.81)$$

A low resistance grounding system is obtained by connecting a small value of the resistance with the grounding conductor or rod. The low resistance grounding system is normally used for the loads which are connected below the 220 V supply and the system is designed to shut down in 10 s. A low resistance grounding system is more expensive than a solidly grounding system. The advantages of low resistance grounding system are lower value of the ground fault current, good control of transient overvoltage, easy fault detection. The disadvantages are high cost and, presence of higher harmonics (in some cases third harmonic component may appear).

4.5.3 Reactance Grounding

The reactance grounding is obtained by connecting a reactor (low inductance) between the neutral point and the ground as shown in Fig. 4.13. The magnitude of the ground fault current can be reduced to an acceptable value by changing the reactance.

Nowadays, this type of grounding is not used due to the requirement of higher value of the fault current in operating protective equipment, and due to the presence of high transient voltage that appears during ground fault condition.

Fig. 4.13 Wye-connection
with reactor grounding

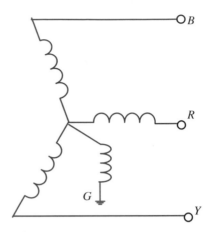

4.5.4 Voltage Transformer Grounding

The voltage transformer grounding is obtained by connecting a neutral of wye connection to the ground through the primary of a single-phase voltage transformer as shown in Fig. 4.14. The secondary of the voltage transformer is connected to the relay through a series resistor.

The primary of the voltage transformer provides high reactance to the neutral of the wye connection. As a result, the neutral point works as an ungrounded system. A voltage is induced at the secondary due to a ground fault to any line of the wye connection. This voltage triggers the operation of the protected equipment. This type of grounding system is used in the generator which is connected to the step-up transformer. The advantages of this grounding system are reduced transient over-voltage and reduced arcing grounds.

Fig. 4.14 Wye connection
with voltage transformer
grounding

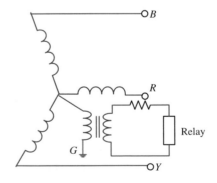

4.6   Resonant Grounding

The resonant grounding is often known as Petersen grounding or arc suppression coil (ASC) grounding. In 1916, W. Petersen was first developed the coil to limit the earth fault current in the ungrounded three-phase system. The concept of resonant grounding comes from the circuit under resonance condition. In this condition, the inductance and the capacitance play an important role. The capacitance is formed in between the energized line and the earth when the transmission lines are passed above the ground. If a specific value of the inductance is connected in parallel with the earth capacitance then the fault current flowing through the inductor will be equal and 180° out of phase to the total capacitor current. The arc suppression coil is adjusted in such a way that the value of the fault current completely balances the total capacitor current, and in this case, it is known resonant grounding. A Petersen coil is an iron core reactor, which is connected to a star-connected three-phase system as shown in Fig. 4.15.

A line to ground fault occurs at phase B. In this condition, the vector diagram is shown in Fig. 4.16. The line B to ground fault voltage is,

$$v_{BG} = 0 \tag{4.82}$$

Applying KVL to the lines Y, B and G yields,

$$v_{YB} + v_{BG} - v_{YG} = 0 \tag{4.83}$$

Substituting Eq. (4.82) into Eq. (4.83) yields,

$$v_{YB} = v_{YG} \tag{4.84}$$

Fig. 4.15 Wye-system with arc suppression coil

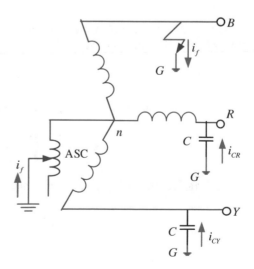

Fig. 4.16 Vector diagram of
a wye-system with faulted line

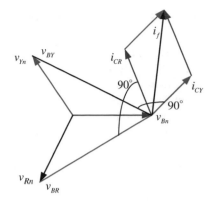

The current in the capacitor due to the line Y is,

$$i_{CY} = \frac{v_{YG}}{X_c} = \frac{v_{YB}}{X_c} = j\omega C\sqrt{3}v_{ph} \tag{4.85}$$

Again, applying KVL to the lines B, G and R yields,

$$v_{BG} - v_{RG} + v_{RB} = 0 \tag{4.86}$$

Substituting Eq. (4.82) into Eq. (4.86) yields,

$$v_{RG} = v_{RB} \tag{4.87}$$

The current in the capacitor due to the line R is,

$$i_{CR} = \frac{v_{RG}}{X_c} = \frac{v_{RB}}{X_c} = \frac{\sqrt{3}v_{ph}}{X_c} \tag{4.88}$$

The vector diagram of voltage and current is shown in Fig. 4.16. From Fig. 4.16, the value of the fault current can be determined as,

$$I_f = 2i_{CR}\cos 30° \tag{4.89}$$

Substituting Eq. (4.88) into Eq. (4.89) yields,

$$i_f = 2 \times \frac{\sqrt{3}v_{ph}}{X_c} \times \frac{\sqrt{3}}{2} = \frac{3v_{ph}}{X_c} \tag{4.90}$$

This fault current is equal to the total capacitive current and it can be expressed as,

$$i_f = i_c = \frac{3v_{ph}}{X_c} \tag{4.91}$$

Again, the fault current is equal to the arc suppression coil current and it can be written as,

$$i_f = i_L = \frac{v_{ph}}{X_L} \tag{4.92}$$

According to the resonance condition, the following relationship can be found,

$$\frac{v_{ph}}{X_L} = \frac{3v_{ph}}{X_c} \tag{4.93}$$

$$\frac{v_{ph}}{\omega L} = \omega C 3 v_{ph} \tag{4.94}$$

$$L = \frac{1}{3\omega^2 C} \tag{4.95}$$

Example 4.2 A Petersen grounding is used in a 66 kV, 50 Hz, 100 km, three-phase transmission line. Each conductor of the transmission line is having a line to earth capacitance of 0.012 μF/km. Determine the value of the inductance and the kVA rating of the Petersen coil when a ground fault occurs in a line.

Solution

The actual value of the line to earth capacitance is,

$$C = 100 \times 0.012 = 1.2 \, \mu F$$

The value of the inductance is,

$$L = \frac{1}{3\omega^2 C} = \frac{1}{3 \times (2\pi \times 50)^2 \times 1.2 \times 10^{-6}} = 2.81 \, H$$

The value of the phase voltage across the Petersen coil is,

$$V_{ph} = \frac{66 \times 1000}{\sqrt{3}} = 38,106.24 \, V$$

The value of the current through the Petersen coil is,

$$I_f = \frac{V_{ph}}{X_L} = \frac{38,106.24}{2\pi \times 50 \times 2.81} = 43.17 \, A$$

The rating of the Petersen coil is,

$$\text{Rating} = V_{ph} I_f = \frac{38,106.24 \times 43.17}{1000} = 1644.89 \, kVA$$

Example 4.3 A 66 kV, 50 Hz, three-phase transmission lines passes near the highway and through mountain. The length of the transmission line is 140 km and the three conductors are arranged vertically and separated from each other by 1.5 m. Consider that the effective radius of the conductor is 0.001 m. Calculate the value of the inductance and the kVA rating of the Petersen coil when a ground fault occurs in one line.

Solution

The value of the capacitance between the line and earth is,

$$C = \frac{2\pi\varepsilon_0}{\ln\frac{d}{r}} = \frac{2\pi \times 8.854 \times 10^{-12}}{\ln\frac{1.5}{0.001}} = 7.61 \times 10^{-12}\,\text{F/m}$$

The actual value of the earth capacitance is,

$$C = 140 \times 1000 \times 7.61 \times 10^{-12} = 1.06 \times 10^{-6}\,\text{F}$$

The value of the Petersen coil is,

$$L = \frac{1}{3\omega^2 C} = \frac{1}{3 \times (2\pi \times 50)^2 \times 1.06 \times 10^{-6}} = 3.19\,\text{H}$$

The value of the phase voltage across the Petersen coil is,

$$V_{ph} = \frac{66 \times 1000}{\sqrt{3}} = 38{,}106.24\,\text{V}$$

The value of the current in the Petersen coil is,

$$I_f = \frac{V_{ph}}{X_L} = \frac{38{,}106.24}{2\pi \times 50 \times 3.19} = 38.02\,\text{A}$$

The rating of the Petersen coil is,

$$\text{Rating} = V_{ph}I_f = \frac{38{,}106.24 \times 38.02}{1000} = 1448.94\,\text{kVA}$$

Practice problem 4.2

A Petersen grounding is used in a 66 kV, 50 Hz, 90 km, three-phase transmission line. Each conductor of a transmission line is having a line to earth capacitance of 0.02 µF/km. Determine the value of the inductive reactance of the Petersen coil.

Practice problem 4.3

A 1200 kVA Petersen coil grounding is used in a 66 kV, 50 Hz, 70 km, three-phase transmission line. Each conductor of the transmission line is having a line to earth capacitance. Find the value of the inductive reactance and the value of the line to earth capacitance.

4.7 Ground Resistance

The ground resistance is often known as earth resistance. The ground resistance is defined as the ratio of voltage across the grounding device and the current flowing into the earth through the grounding device.

The value of the ground resistance is the combination of grounding device resistance, lead resistance, contact resistance between grounding device and soil, and surrounding soil resistance. The surrounding soil resistance is normally higher compared to other components. However, the value of the earth resistance, mainly depends on the shape of the grounding device, type of the soil and wetness of the soil. The properties of the soil changes within the small areas of the earth. The model of the ground resistance is shown in Fig. 4.17.

Fig. 4.17 Model of ground resistance

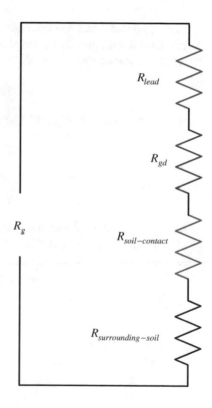

4.8   Electric Potential

An electric potential is an essential component which is used to calculate the ground resistance for a given injected current in the ground. Consider that a test charge q is placed at a point A as shown in Fig. 4.18. This charge needs to be moved from the initial point A to the final point B. A force $\mathbf{F}$ is directed from the test charge to the right direction and a force $\mathbf{F}'$ is directed to the left direction as shown in Fig. 4.18. At point A, the following equation can be written when the charge is not moving anywhere,

$$\mathbf{F} = -\mathbf{F}' \tag{4.96}$$

In this case, the electric field intensity, which is defined as the force per unit charge, it can be written as,

$$\mathbf{E} = \frac{\mathbf{F}}{q} \tag{4.97}$$

Again, consider that the test charge q moves slowly to a small distance $d\mathbf{l}$ towards point B with the application of force $\mathbf{F}'$ as shown in Fig. 4.19. The work done to move the test charge for the distance $d\mathbf{l}$ is,

$$dW = \mathbf{F}'.d\mathbf{l} \tag{4.98}$$

Substituting Eqs. (4.96) and (4.97) into Eq. (4.98) yields,

$$dW = -q\mathbf{E}.d\mathbf{l} \tag{4.99}$$

$$\frac{dW}{q} = -\mathbf{E}.d\mathbf{l} \tag{4.100}$$

Consider that the difference in electric potential between the points of the path $d\mathbf{l}$ is represented by dV, and this can be written as,

$$dV = V_B - V_A = \frac{dW}{q} \tag{4.101}$$

Fig. 4.18 Test charge with electric field

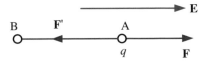

Fig. 4.19 Test charge with a small distance

Substituting Eq. (4.100) into Eq. (4.101) yields,

$$dV = -\mathbf{E}.d\mathbf{l} \tag{4.102}$$

Integrating Eq. (4.102) yields,

$$\int_{initial}^{final} dV = -\int_{initial}^{final} \mathbf{E}.d\mathbf{l} \tag{4.103}$$

$$\int_{A}^{B} dV = -\int_{A}^{B} \mathbf{E}.d\mathbf{l} \tag{4.104}$$

$$V_B - V_A = -\int_{A}^{B} \mathbf{E}.d\mathbf{l} \tag{4.105}$$

$$V_A - V_B = V_{AB} = \int_{A}^{B} \mathbf{E}.d\mathbf{l} \tag{4.106}$$

4.9   Ground Resistance with Hemisphere

Figure 4.20 shows a hemisphere buried in the soil whose resistivity is ρ. The hemisphere is connected to the conductor. The current I in the conductor passes to the hemisphere and distributed to the ground. The current density at the surface of the hemisphere with surface area A is,

$$J = \frac{I}{A} = \frac{I}{2\pi r^2} \tag{4.107}$$

where r is the radius of the hemisphere. The current density at any distance x from the center of the hemisphere is,

$$J(x) = \frac{I}{2\pi x^2} \tag{4.108}$$

Fig. 4.20 Hemisphere with
current lines

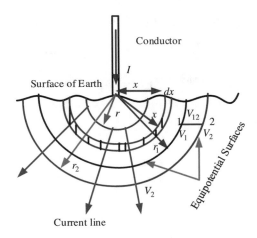

The electric field at any distance x from the center of the hemisphere can be determined as,

$$J(x) = \sigma E(x) \tag{4.109}$$

Substituting Eq. (4.108) into Eq. (4.109) yields,

$$\frac{I}{2\pi x^2} = \sigma E(x) \tag{4.110}$$

$$\frac{I}{\sigma 2\pi x^2} = E(x) \tag{4.111}$$

$$E(x) = \frac{\rho I}{2\pi x^2} \tag{4.112}$$

The electric potential for a line is defined as,

$$V = -\int_{initial}^{final} \mathbf{E} \cdot d\mathbf{l} \tag{4.113}$$

Inside the earth, the potential difference between the two points can be calculated as,

$$V_{21} = V_2 - V_1 = -\int_{x=r_1}^{x=r_2} E(x) dx \tag{4.114}$$

Substituting Eq. (4.112) into Eq. (4.114) yields,

$$V_{12} = \int_{x=r_1}^{x=r_2} \frac{\rho I}{2\pi x^2} dx \tag{4.115}$$

$$V_{12} = V_1 - V_2 = -\frac{\rho I}{2\pi} \left[\frac{1}{x}\right]_{r_1}^{r_2} \tag{4.116}$$

$$V_{12} = -\frac{\rho I}{2\pi} \left[\frac{1}{r_2} - \frac{1}{r_1}\right] \tag{4.117}$$

$$V_{12} = \frac{\rho I}{2\pi} \left[\frac{1}{r_1} - \frac{1}{r_2}\right] \tag{4.118}$$

If $r_1 = r$ and $r_2 = \infty$, then the expression of electric potential between the two points can be reduced as,

$$V_{12} = \frac{\rho I}{2\pi} \left[\frac{1}{r} - \frac{1}{\infty}\right] \tag{4.119}$$

$$V_{12} = \frac{\rho I}{2\pi r} \tag{4.120}$$

Then, the expression of ground resistance can be calculated as,

$$R_g = \frac{V_{12}}{I} = \frac{\rho}{2\pi r} \tag{4.121}$$

Example 4.4 A house owner has buried a 1.5 m radius hemisphere earth electrode into his neighboring soil whose resistivity is found to be 20 Ω-m. Determine the ground resistance of the house owner. In addition, find the ground resistance at a distance of 10 m from the center of the hemisphere.

Solution

The value of the ground resistance of the house owner can be determined as,

$$R_g = \frac{\rho}{2\pi r} = \frac{20}{2\pi \times 1.5} = 2.12 \, \Omega$$

The value of the ground resistance at a distance of 10 m can be calculated as,

$$R_g = \frac{\rho}{2\pi}\left[\frac{1}{r_1} - \frac{1}{r_2}\right] = \frac{20}{2\pi}\left(\frac{1}{1.5} - \frac{1}{10}\right) = 1.80\,\Omega$$

Practice problem 4.4

The ground resistance of a house owner has found to be 0.6 Ω using a hemisphere earth electrode. The resistivity of the soil is found to be 10 Ω-m. Calculate the radius of the hemisphere.

4.10   Ground Resistance with Sphere Electrode

A sphere with a radius r is shown in Fig. 4.21. The surface area of a sphere can be written as,

$$A_{ss} = 4\pi r^2 \tag{4.122}$$

For the current I, the expression of current density is,

$$J_s = \frac{I}{A_{sp}} \tag{4.123}$$

Substituting Eq. (4.122) into Eq. (4.123) yields,

$$J_s = \frac{I}{4\pi r^2} \tag{4.124}$$

Fig. 4.21 Sphere with current lines

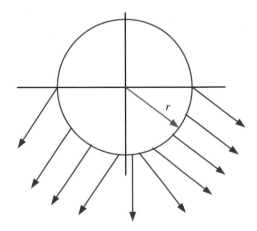

The current density of the sphere at any distance $x \geq r$ is,

$$J_s(x) = \frac{I}{4\pi x^2} \tag{4.125}$$

The electric field of the sphere at any distance $x \geq r$ is,

$$E_{sp}(x) = \frac{\rho I}{4\pi x^2} \tag{4.126}$$

The potential from the distance r to infinity can be determined as,

$$V_{sp} = \int_{x=r}^{x=\infty} \frac{\rho I}{4\pi x^2} dx \tag{4.127}$$

$$V_{sp} = -\frac{\rho I}{4\pi} \left[\frac{1}{x} \right]_{x=r}^{x=\infty} \tag{4.128}$$

$$V_{sp} = -\frac{\rho I}{4\pi} \left[\frac{1}{\infty} - \frac{1}{r} \right] \tag{4.129}$$

$$V_{sp} = \frac{\rho I}{4\pi r} \tag{4.130}$$

Then, the expression of ground resistance of the sphere can be calculated as,

$$R_g = \frac{V_{sp}}{I} = \frac{\rho}{4\pi r} \tag{4.131}$$

Example 4.5 A 1 m radius sphere earth electrode is used for a grounding system at a house and the soil resistivity at that place is found to be 30 Ω-m. Calculate the ground resistance.

Solution

The value of the ground resistance can be determined as,

$$R_g = \frac{\rho}{4\pi r} = \frac{30}{4\pi \times 1} = 2.38 \, \Omega$$

Practice problem 4.5
A sphere earth electrode is used for a grounding system at a house. The soil resistivity and the ground resistance at that place are found to be 30 Ω-m and 1.5 Ω, respectively. Determine the radius of the sphere earth electrode.

4.11 Ground Resistance with Cylindrical Rod

A cylindrical rod with a radius r and length l is inserted vertically into the soil. A current I is passed into the rod from the top end as shown in Fig. 4.22, and this current is dividing the surface into two parts. The first part is due to the current coming out horizontally along the length of the cylindrical rod, forming a cylindrical surface of radius x and length l. The second part is due to the current coming out radially from the bottom of the cylindrical rod, forming a hemisphere of radius x. In this case, the surface area of the rod at any radial distance x from the longitudinal axis of the cylinder can be written as,

$$A = \pi x^2 + \pi x^2 + 2\pi x l \tag{4.132}$$

At any radial distance x, the electric field can be written as,

$$E(x) = \rho J(x) = \rho \frac{I}{A} \tag{4.133}$$

Substituting Eq. (4.132) into Eq. (4.133) yields,

$$E(x) = \rho \frac{I}{2\pi x^2 + 2\pi x l} \tag{4.134}$$

Fig. 4.22 Cylindrical rod conductor

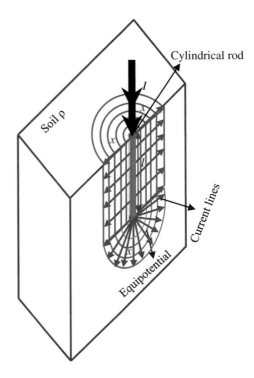

The voltage from the distance r to infinity can be derived as,

$$V = \frac{\rho I}{2\pi} \int_r^\infty \frac{1}{x(l+x)} dx \tag{4.135}$$

We need to apply partial fraction decomposition as,

$$\frac{1}{x(l+x)} = \frac{A}{x} + \frac{B}{(l+x)} \tag{4.136}$$

$$1 = A(l+x) + Bx \tag{4.137}$$

$$A = \frac{1}{l}, \quad B = -\frac{1}{l} \tag{4.138}$$

Then Eq. (4.135) can be modified as,

$$V = \frac{\rho I}{2\pi} \left(\int_r^\infty \frac{1}{lx} dx - \int_r^\infty \frac{1}{l(x+l)} dx \right) \tag{4.139}$$

$$V = \frac{\rho I}{2\pi l} [\ln x - \ln(x+l)]_r^\infty \tag{4.140}$$

$$V = \frac{\rho I}{2\pi l} \left[\ln \frac{x}{x+l} \right]_r^\infty \tag{4.141}$$

$$V = \frac{\rho I}{2\pi l} \left[0 - \ln \frac{r}{r+l} \right] \tag{4.142}$$

$$V = \frac{\rho I}{2\pi l} \left[\ln \frac{r+l}{r} \right] \tag{4.143}$$

$$R_g = \frac{\rho}{2\pi l} \ln \frac{r+l}{r} \tag{4.144}$$

Assuming that the radius of the cylindrical rod is very small compared to its length, the expression of the ground resistance can be written as,

$$R_g = \frac{\rho}{2\pi l} \ln \frac{l}{r} \tag{4.145}$$

If the total length of the cylindrical rod is $2l$, then the expression of the ground resistance becomes,

$$R_g = \frac{\rho}{2\pi l} \ln \frac{2l}{r} \qquad (4.146)$$

Again, substituting $r = \frac{d}{2}$ in Eq. (4.146) yields,

$$R_g = \frac{\rho}{2\pi l} \ln \frac{4l}{d} \qquad (4.147)$$

Example 4.6 A 0.5 in. diameter and 10 ft length cylindrical electrode is driven vertically into the silt soil for the grounding system. The resistivity of the silt soil is found to be 25 Ω-ft. Calculate the ground resistance.

Solution

The value of the ground resistance can be determined as,

$$R_g = \frac{\rho}{2\pi l} \ln \frac{2l}{r} = \frac{25}{2\pi \times 10} \ln \frac{20}{0.25} = 1.74\ \Omega$$

Practice problem 4.6
The ground resistance of a 1 in. diameter and 8 ft cylindrical electrode for the clay soil is found to be 1.5 Ω. Calculate the resistivity of the clay soil.

Alternative approach:
In this case, the cylindrical rod is driven vertically into the soil as shown in Fig. 4.23. Consider that the equipotential surface due to the current coming out of the soil has a shape of a prolate ellipsoid, which minor axis is very small compared to the major axis as shown in Fig. 4.24.

The electrostatic capacity (C) of the prolate ellipsoid in revolution is given by A. Gray as,

$$C = \frac{\sqrt{c^2 - a^2}}{\ln \frac{\sqrt{c^2 - a^2} + c}{a}} \qquad (4.148)$$

where c is the length of the major semi-axis and a is the length of the minor semi-axis.

Equation (4.148) can be written as,

$$C = \frac{c\sqrt{1 - \frac{a^2}{c^2}}}{\ln \frac{c\sqrt{1 - \frac{a^2}{c^2}} + c}{a}} \qquad (4.149)$$

Fig. 4.23 Cross-section of
the soil with a buried
cylindrical rod electrode

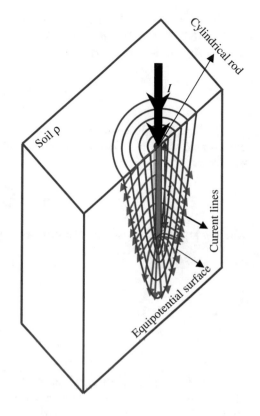

Fig. 4.24 Prolate ellipsoid

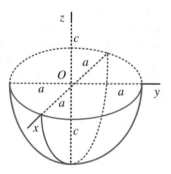

Since c is very large compared to a, the value of a^2/c^2 is almost zero and can be neglected. Equation (4.149) can be reduced as,

$$C = \frac{c}{\ln \frac{2c}{a}} \tag{4.150}$$

Now, if we replace the length of the major semi-axis (C) with the length of the vertical rod (l) and length of the minor semi-axis (a) with the radius r, then Eq. (4.150) can be written as,

$$C = \frac{l}{\ln \frac{2l}{r}} \tag{4.151}$$

C.F. Tagg introduced an equation to obtain the ground resistance (R_g) from the electrostatic capacity (C) for any electrode of any shape, i.e. sphere, plate and rod, as follows:

$$R_g = \frac{\rho}{2\pi C} \tag{4.152}$$

where ρ is the soil resistivity. Substituting Eq. (4.151) into Eq. (4.152) yields,

$$R_g = \frac{\rho}{2\pi \dfrac{l}{\ln \dfrac{2l}{r}}} \tag{4.153}$$

$$R_g = \frac{\rho}{2\pi l} \ln \frac{2l}{r} \tag{4.154}$$

The method used by H.B. Dwight and G.F. Tagg to derive the ground resistance of a cylindrical rod is similar to deriving the electrostatic capacity of an isolated cylinder. Therefore, the problem of deriving the ground resistance is reduced to obtaining the electrostatic capacity of the cylinder. First, we need to consider that the cylindrical rod is buried horizontally into the ground whose length is $2l$ and the diameter is $2a$ as shown in Fig. 4.25.

The cylindrical rod has a uniform charge density along its length so that the charge on the ring is the value of the total charge q per length times the length of the ring:

$$l_r = qdy \tag{4.155}$$

The potential at point P in Fig. 4.26 is calculated as the charge on the ring over the distance between point P and the center of the ring as,

$$\varphi = \frac{qdy}{\sqrt{a^2 + y^2}} \tag{4.156}$$

The electrostatic capacity is the total charge $q \times 2l$ of the cylindrical rod over its average potential. The average potential of the rod can be obtained by integrating

Fig. 4.25 Cross-section of
the soil with a buried
cylindrical rod electrode

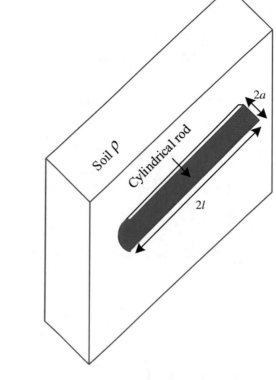

Fig. 4.26 Cylindrical rod
inserted into the soil in two
dimensions

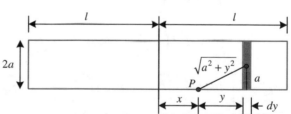

twice the potential at point P from $y = 0$ to $y = l - x$ and from $y = 0$ to $y = l + x$.
These are,

$$V_1 = \int_0^{l-x} \frac{q\,dy}{\sqrt{a^2 + y^2}} = q\left[\sinh^{-1}\left(\frac{y}{a}\right)\right]_0^{l-x} = q\left[\sinh^{-1}\left(\frac{l-x}{a}\right)\right] \qquad (4.157)$$

$$V_2 = \int_0^{l+x} \frac{q\,dy}{\sqrt{a^2 + y^2}} = q\left[\sinh^{-1}\left(\frac{y}{a}\right)\right]_0^{l+x} = q\left[\sinh^{-1}\left(\frac{l+x}{a}\right)\right] \qquad (4.158)$$

Then, after multiplying Eqs. (4.157) and (4.158) multiply by $\frac{dx}{l}$ and integrate from $x = 0$ to $x = l$ as follows:

$$I_1 = \int_0^l \frac{q}{l} \sinh^{-1}\left(\frac{l-x}{a}\right) dx$$

$$= \frac{q}{l}\left[l \sinh^{-1}\left(\frac{x-l}{a}\right) - x \sinh^{-1}\left(\frac{x-l}{a}\right) + \sqrt{(x-l)^2 + a^2} \right]_0^l \qquad (4.159)$$

$$I_1 = \frac{q}{l}\left[l \sinh^{-1}\left(\frac{l-l}{a}\right) - l \sinh^{-1}\left(\frac{l-l}{a}\right) + \sqrt{(l-l)^2 + a^2} - l \sinh^{-1}\left(\frac{0-l}{a}\right) \right.$$
$$\left. + 0 \sinh^{-1}\left(\frac{0-l}{a}\right) - \sqrt{(0-l)^2 + a^2} \right]$$

$$\qquad (4.160)$$

$$I_1 = \frac{q}{l}\left[l \sinh^{-1}(0) - l \sinh^{-1}(0) + \sqrt{(0)^2 + a^2} - l \sinh^{-1}\left(\frac{-l}{a}\right) + 0 \sinh^{-1}\left(\frac{-l}{a}\right) - \sqrt{l^2 + a^2} \right]$$

$$\qquad (4.161)$$

$$I_1 = \frac{q}{l}\left[a + l \sinh^{-1}\left(\frac{l}{a}\right) - \sqrt{l^2 + a^2} \right] \qquad (4.162)$$

$$I_2 = \int_0^l \frac{q}{l} \sinh^{-1}\left(\frac{l+x}{a}\right) dx$$

$$= \frac{q}{l}\left[l \sinh^{-1}\left(\frac{x+l}{a}\right) + x \sinh^{-1}\left(\frac{x+l}{a}\right) - \sqrt{(x+l)^2 + a^2} \right]_0^l \qquad (4.163)$$

$$I_2 = \frac{q}{l}\left[l \sinh^{-1}\left(\frac{l+l}{a}\right) + l \sinh^{-1}\left(\frac{l+l}{a}\right) - \sqrt{(l+l)^2 + a^2} - l \sinh^{-1}\left(\frac{0+l}{a}\right) \right.$$
$$\left. - 0 \sinh^{-1}\left(\frac{0+l}{a}\right) + \sqrt{(0+l)^2 + a^2} \right]$$

$$\qquad (4.164)$$

$$I_2 = \frac{q}{l}\left[l \sinh^{-1}\left(\frac{2l}{a}\right) + l \sinh^{-1}\left(\frac{2l}{a}\right) - \sqrt{4l^2 + a^2} - l \sinh^{-1}\left(\frac{l}{a}\right) - 0 + \sqrt{l^2 + a^2} \right]$$

$$\qquad (4.165)$$

$$I_2 = \frac{q}{l}\left[2l \sinh^{-1}\left(\frac{2l}{a}\right) - \sqrt{4l^2 + a^2} - l \sinh^{-1}\left(\frac{l}{a}\right) + \sqrt{l^2 + a^2} \right] \qquad (4.166)$$

The total average potential can be determined by summing Eqs. (4.162) and (4.166) yields,

$$V_{avg} = \frac{q}{l}\left[a + l\sinh^{-1}\left(\frac{l}{a}\right) - \sqrt{l^2 + a^2} + 2l\sinh^{-1}\left(\frac{2l}{a}\right) - \sqrt{4l^2 + a^2}\right.$$
$$\left. -l\sinh^{-1}\left(\frac{l}{a}\right) + \sqrt{l^2 + a^2}\right] \tag{4.167}$$

$$V_{avg} = \frac{q}{l}\left[a + 2l\sinh^{-1}\left(\frac{2l}{a}\right) - \sqrt{4l^2 + a^2}\right] \tag{4.168}$$

The inverse hyperbolic sine can be written by using natural logarithm as follows,

$$\sinh^{-1}(x) = \ln\left[x + \sqrt{x^2 + 1}\right] \tag{4.169}$$

Then, Eq. (4.168) can be expressed as,

$$V_{avg} = \frac{q}{l}\left[a + 2l\ln\left[\frac{2l}{a} + \sqrt{\left(\frac{2l}{a}\right)^2 + 1}\right] - \sqrt{4l^2 + a^2}\right] \tag{4.170}$$

$$V_{avg} = \frac{q}{l}\left[a + 2l\ln\left[\frac{2l}{a} + \sqrt{\frac{4l^2}{a^2} + 1}\right] - \sqrt{4l^2 + a^2}\right] \tag{4.171}$$

$$V_{avg} = \frac{q}{l}\left[a + 2l\ln\left[\frac{2l}{a} + \frac{2l}{a}\sqrt{1 + \frac{a^2}{4l^2}}\right] - 2l\sqrt{1 + \frac{a^2}{4l^2}}\right] \tag{4.172}$$

$$V_{avg} = \frac{q}{l}(2l)\left[\ln\left[\frac{2l}{a} + \frac{2l}{a}\sqrt{1 + \frac{a^2}{4l^2}}\right] + \frac{a}{2l} - \sqrt{1 + \frac{a^2}{4l^2}}\right] \tag{4.173}$$

$$V_{avg} = 2q\left[\ln\left[\frac{2l}{a} + \frac{2l}{a}\sqrt{1 + \left(\frac{a}{2l}\right)^2}\right] + \frac{a}{2l} - \sqrt{1 + \left(\frac{a}{2l}\right)^2}\right] \tag{4.174}$$

Since, the radius of the rod is very small compared to the length of the rod, the term $\frac{a}{2l}$ goes to zero. Then, Eq. (4.174) can be reduced as,

$$V_{avg} = 2q\left[\ln\left(\frac{2l}{a} + \frac{2l}{a}\right) - 1\right] \tag{4.175}$$

$$V_{avg} = 2q\left[\ln\left(\frac{4l}{a}\right) - 1\right] \tag{4.176}$$

Then, the electrostatic capacity of the cylindrical rod can be determined from the following expression,

$$C = \frac{2lq}{V_{avg}} \tag{4.177}$$

Substituting Eq. (4.176) into Eq. (4.177) yields,

$$C = \frac{2lq}{V_{avg}} = \frac{2ql}{2q\left[\ln\left(\frac{4l}{a}\right) - 1\right]} \tag{4.178}$$

$$C = \frac{l}{\left[\ln\left(\frac{4l}{a}\right) - 1\right]} \tag{4.179}$$

Again, substituting Eq. (4.179) into Eq. (4.152) yields the expression of ground resistance,

$$R_g = \frac{\rho}{2\pi \frac{l}{\left(\ln\frac{4l}{a} - 1\right)}} \tag{4.180}$$

$$R_g = \frac{\rho}{2\pi l}\left(\ln\frac{4l}{a} - 1\right) \tag{4.181}$$

Substituting $a = \frac{d}{2}$ in Eq. (4.181) yields,

$$R_g = \frac{\rho}{2\pi l}\left(\ln\frac{8l}{d} - 1\right) \tag{4.182}$$

Example 4.7 A 1 in. diameter and 8 ft length cylindrical electrode is driven horizontally into silt soil for the grounding system. The resistivity of the silt soil is found to be 25 Ω-ft. Calculate the ground resistance by using G.F. Tagg formula.

Solution

The value of the ground resistance can be determined as,

$$R_g = \frac{\rho}{2\pi l}\left(\ln\frac{8l}{d} - 1\right) = \frac{25}{2\pi \times 4}\left(\ln\frac{8 \times 4}{1} - 1\right) = 2.45\,\Omega$$

Practice problem 4.7
The ground resistance of a 0.5 in. diameter and 20 ft cylindrical electrode for the sandy soil is found to be 4.5 Ω. Calculate the resistivity of the sandy soil using the G.F. Tagg formula.

4.12   Ground Resistance with Circular Plate

A circular plate is having a radius r inserted vertically into the soil. A cross-section of the soil with a buried circular plate electrode is shown in Fig. 4.27. The current lines coming out of the circular plate have two different directions. The current lines coming out from the edges are radial and create a surface similar to a quarter of a torus, whereas the current lines coming out of the bottom of the circular plate are vertical which form a circumference of radius r. The internal radius of such torus is r and the external radius is x as shown in Fig. 4.27. A two dimensional representation of a circular plate inserted into the soil is shown in Fig. 4.28. In this case, we need to find the area of a part of the torus by using the first theorem of Pappus. This theorem says that the surface area A of a surface of revolution generated by the revolution of a curve about an external axis is equal to the product of the perimeter p of the generating curve and the distance d travelled by the curve's geometric centroid x_c. Therefore, the formula for the surface area is given as,

$$A = p \times d = p \times 2\pi x_c \tag{4.183}$$

The quarter circular arc and its geometry is shown in Fig. 4.29. The radius of the arc is considered as a. The centroid of the quarter-circular arc can be determined as,

$$x_c = \frac{1}{\frac{\pi a}{2}} \int_0^{-\frac{\pi}{2}} y a d\theta \tag{4.184}$$

Fig. 4.27 Cross-section of the soil with a buried circular plate electrode

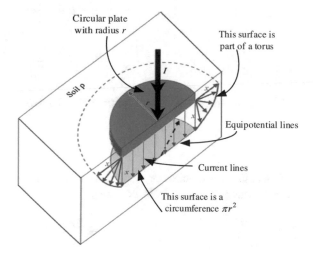

Circular plate with radius r

This surface is part of a torus

Soil ρ

I

Equipotential lines

Current lines

This surface is a circumference πr^2

Fig. 4.28 Circular plate inserted into the soil in two dimensions

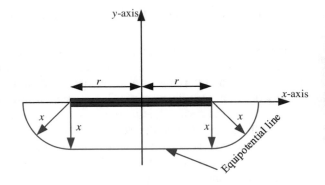

Fig. 4.29 Quarter-circular arc and its geometric centroid

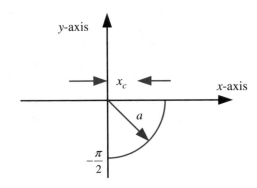

$$x_c = \frac{2}{\pi a} \int_0^{-\frac{\pi}{2}} (a \sin \theta) a d\theta \tag{4.185}$$

$$x_c = \frac{2a^2}{\pi a}(1) = \frac{2a}{\pi} \tag{4.186}$$

The geometric centroid of a quarter-circular arc of radius x is,

$$x_c = \frac{2x}{\pi} \tag{4.187}$$

Since, the quarter-circular arc is displaced from the external axis by a distance r, the new value of the centroid becomes,

$$x_c = \frac{2x}{\pi} + r \tag{4.188}$$

The perimeter of the arc is,

$$p = \frac{x\pi}{2} \tag{4.189}$$

Substituting Eqs. (4.188) and (4.189) into Eq. (4.183), the area in revolution yields,

$$A = \frac{\pi x}{2}\left[2\pi\left(\frac{2x}{\pi} + r\right)\right] = \frac{\pi x}{2}[4x + 2\pi r] \tag{4.190}$$

$$A = 2\pi x^2 + \pi^2 r x \tag{4.191}$$

Finally, the complete area in revolution is the summation of the area of part of the torus and the area of the circumference of radius r, which is,

$$A(x) = 2\pi x^2 + \pi^2 r x + \pi r^2 \tag{4.192}$$

The total current density throughout the soil at any distance x is:

$$J(x) = \frac{I}{A(x)} \tag{4.193}$$

Substituting Eq. (4.192) into Eq. (4.193) yields,

$$J(x) = \frac{I}{2\pi x^2 + \pi^2 r x + \pi r^2} \tag{4.194}$$

The electric field at any distance x is,

$$E(x) = \rho J(x) \tag{4.195}$$

Substituting Eq. (4.194) into Eq. (4.195) yields,

$$E(x) = \frac{\rho I}{2\pi x^2 + \pi^2 r x + \pi r^2} \tag{4.196}$$

The potential difference between the center of the circular plate and the remote earth is,

$$V_{0\infty} = \int_0^\infty E(x)dx \tag{4.197}$$

Substituting Eq. (4.196) into Eq. (4.197) yields,

$$V_{0\infty} = \int_0^\infty \frac{\rho I}{2\pi x^2 + \pi^2 rx + \pi r^2} dx \qquad (4.198)$$

$$\frac{1}{2\pi x^2 + \pi r^2 + \pi^2 rx} = \frac{1}{6.28x^2 + 3.14r^2 + 9.87rx} = \frac{0.16}{x^2 + 1.57rx + 0.5r^2} \qquad (4.199)$$

$$\frac{1}{(x+1.13r)(x+0.45r)} = \frac{-1.47}{r(x+1.13r)} + \frac{1.47}{r(x+0.45r)} \qquad (4.200)$$

Equation (4.198) can be modified as,

$$V_{0\infty} = \frac{0.16 \times 1.47}{r} \rho I \int_{x=0}^{x=\infty} \left[\frac{1}{(x+0.45r)} - \frac{1}{(x+1.13r)} \right] dx \qquad (4.201)$$

$$V_{0\infty} = \frac{0.24}{r} \rho I \ln \left(\frac{x+0.45r}{x+1.13r} \right)_0^\infty \qquad (4.202)$$

$$V_{0\infty} = \frac{0.24}{r} \rho I [\ln(\infty + 0.45r) - \ln(\infty + 1.13r) - \ln(0.45r) + \ln(1.13r)] \qquad (4.203)$$

$$V_{0\infty} = \frac{0.24}{r} \rho I \left[\ln \left(\frac{1.13r}{0.45r} \right) \right] \qquad (4.204)$$

$$V_{0\infty} = \frac{\rho I}{4.5r} \qquad (4.205)$$

The expression of the ground resistance can be written as,

$$R_g = \frac{\rho}{4.5r} \qquad (4.206)$$

Many assumptions have been taken during derivation process of the ground resistance. Therefore, the general expression of the ground resistance is,

$$R_g = \frac{\rho}{4r} \qquad (4.207)$$

Now, let's discuss a real life example on ground resistance. Let's assume that a person is standing near the fence of an electrical substation in bare feet. In this case, the person will have a resistance of each foot and the center of the earth as shown as shown in Fig. 4.30. The resistances due to hands and upper body parts such as lungs

Fig. 4.30 Person with feet
resistance

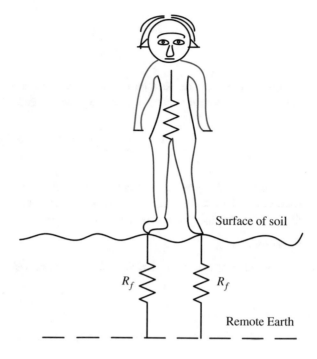

Surface of soil

R_f R_f

Remote Earth

will be neglected due to an open circuit. Consider that the resistance between the bottom of the foot and the center of the earth is R_f. Assume that the area of the footprint is the same as the plate area. According to ground resistance of circular plate as derived in Eq. (4.169), the ground resistance of a single foot will be,

$$R_f = \frac{\rho}{4r} \tag{4.208}$$

The area of the circulate plate is,

$$A = \pi r^2 \tag{4.209}$$

$$r = \sqrt{\frac{A}{\pi}} \tag{4.210}$$

Substituting Eq. (4.210) into Eq. (4.208) provides,

$$R_f = \frac{\rho}{4.5\sqrt{\frac{A}{\pi}}} \tag{4.211}$$

Consider an approximate area of footprint is equal to 0.02 m^2, then the expression of ground resistance becomes,

$$R_f = \frac{\rho}{4\sqrt{\frac{0.02}{\pi}}} = 3.13\,\rho \approx 3\,\rho \qquad (4.212)$$

The total ground resistance of a person will be the parallel combination of two single foot resistance values, and the expression becomes,

$$R_g = \frac{R_f \times R_f}{R_f + R_f} = 0.5\,R_f = 1.5\,\rho \qquad (4.213)$$

Example 4.8 The ground resistance of a 12 in. diameter circular plate electrode is found to be 2 Ω in the garden soil. A person is standing on the soil surface in bare feet. Determine the ground resistance of the person.

Solution

The value of the soil resistivity can be determined as,

$$R_g = \frac{\rho}{4r}$$

$$\rho = 4rR_g = 4 \times 0.5 \times 2 = 4\,\Omega\text{-ft}$$

The value of the ground resistance can be calculated as

$$R_g = 1.5\,\rho = 1.5 \times 4 = 6\,\Omega$$

Practice problem 4.8
The ground resistance of a person is found to be 4 Ω when standing on the soil surface in bare feet. The ground resistance obtained using the circular plate electrode is 2.5 Ω. Calculate the radius of the circular plate electrode.

4.13 Ground Resistance with Conductor Type Electrode

In the grid system, sometimes both conductors and rods are used to get an acceptable value of the ground resistance at the substation. The L- and I-types connectors are usually used to make a grid. Figure 4.31 shows a conductor electrode partially buried into the soil. The conductor type electrode I similar to the cylindrical rod type electrode [1, 2]. At any radial distance x, consider the half cylindrical rod with a hemisphere at both ends. In this condition, the surface area is $\pi x^2 + \pi x^2$.

In this case, we have a cylindrical conductor half-buried into the soil as shown in Fig. 4.31. We can divide the surface in three parts. The first part is due to the current coming out horizontally along the length of the cylindrical conductor, forming a semi-cylindrical surface of radius x and length l. The second part is due to the

Fig. 4.31 Cross-section of
the soil with a half-buried
cylindrical conductor type
electrode

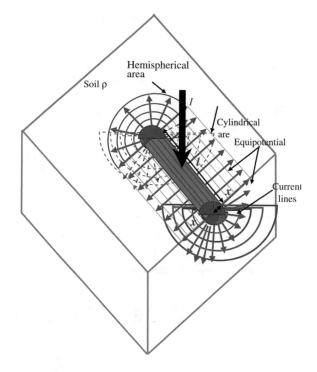

current coming out radially of the one of the ends of the cylindrical conductor,
forming half of a hemisphere of radius x. The third part is the same as the second
part.

From Fig. 4.31, the total surface area can be calculated as,

$$A(x) = \pi x l + \pi x^2 + \pi x^2 \tag{4.214}$$

The total current density throughout the soil at any distance x is,

$$J(x) = \frac{I}{A(x)} = \frac{I}{\pi x l + \pi x^2 + \pi x^2} \tag{4.215}$$

The electric field at any distance x is,

$$E(x) = \rho J(x) \tag{4.216}$$

Substituting Eq. (4.215) into Eq. (4.216) yields,

$$E(x) = \frac{\rho I}{\pi x l + \pi x^2 + \pi x^2} \tag{4.217}$$

The potential difference between the conductor and the remote earth is [3, 4],

$$V = \int_r^\infty E(x)dx \tag{4.218}$$

Substituting Eq. (4.217) into Eq. (4.218) provides,

$$V = \int_r^\infty \frac{\rho I}{\pi x l + 2\pi x^2} dx \tag{4.219}$$

$$V = \frac{\rho I}{2\pi} \int_r^\infty \frac{1}{x\left(\frac{l}{2}+x\right)} dx \tag{4.220}$$

We need to apply partial fraction decomposition in the following ways,

$$\frac{1}{x\left(\frac{l}{2}+x\right)} = \frac{A}{x} + \frac{B}{\left(\frac{l}{2}+x\right)} \tag{4.221}$$

$$\frac{1}{x\left(\frac{l}{2}+x\right)} = \frac{A\left(\frac{l}{2}+x\right) + Bx}{x\left(\frac{l}{2}+x\right)} \tag{4.222}$$

$$1 = A\left(\frac{l}{2}+x\right) + Bx \tag{4.223}$$

Substituting $x = 0$ into Eq. (4.223) yields,

$$A = \frac{2}{l} \tag{4.224}$$

Again, substituting $x = -\frac{l}{2}$ into Eq. (4.223) yields,

$$B = -\frac{2}{l} \tag{4.225}$$

Equation (4.220) can be modified as,

$$V = \frac{\rho I}{2\pi}\left[\int_r^\infty \frac{2}{lx}dx - \int_r^\infty \frac{2}{l\left(x+\frac{l}{2}\right)}dx\right] \tag{4.226}$$

$$V = \frac{\rho I}{2\pi l} \left[\int_{r}^{\infty} \frac{2}{x} dx - \int_{r}^{\infty} \frac{2}{x + \frac{l}{2}} dx \right] \tag{4.227}$$

$$V = \frac{\rho I}{\pi l} \left[\ln x - \ln\left(x + \frac{l}{2}\right) \right]_{r}^{\infty} \tag{4.228}$$

$$V = \frac{\rho I}{\pi l} \left[\ln \frac{x}{x + \frac{l}{2}} \right]_{r}^{\infty} \tag{4.229}$$

$$V = \frac{\rho I}{\pi l} \left[\ln \frac{x}{\frac{2x+l}{2}} \right]_{r}^{\infty} \tag{4.230}$$

$$V = \frac{\rho I}{\pi l} \left[\ln \frac{2x}{2x + l} \right]_{r}^{\infty} \tag{4.231}$$

$$V = \frac{\rho I}{\pi l} \left[0 - \ln \frac{2r}{2r + l} \right] \tag{4.232}$$

$$V = \frac{\rho I}{\pi l} \left[\ln \frac{2r + l}{2r} \right] \tag{4.233}$$

Finally, the ground resistance of the conductor electrode can be expressed as,

$$R_g = \frac{V}{I} = \frac{\rho}{\pi l} \left[\ln \frac{2r + l}{2r} \right] \tag{4.234}$$

Substituting $d = 2r$ in (4.234) yields,

$$R_g = \frac{\rho}{\pi l} \ln \frac{d + l}{d} \tag{4.235}$$

Example 4.9 A 0.5 in. diameter and 16 ft length of conductor type electrode is used for the grounding system at the sandy soil. The resistivity of the sandy soil is given by 140 Ω-ft. Find ground resistance.

Solution

The value of the soil resistivity can be determined as,

$$R_g = \frac{\rho}{\pi l} \ln \frac{d + l}{d} = \frac{140}{\pi \times 16} \ln \frac{\frac{0.5}{12} + 16}{\frac{0.5}{12}} = 18.51\,\Omega$$

Practice problem 4.9

A 1 in. diameter and 10 ft length of conductor type electrode is used for a residential grounding system. Find the resistivity of the soil if the value of the ground resistance is given by 0.9 Ω.

Exercise Problems

4.1 A 2 MVA, 11/0.415 kV delta-wye three-phase transformer is connected to the electrical network of a liquefy gas company. The line-to ground fault occurs in the secondary of the transformer. The transformer impedance and the ground fault impedance are found to be 8 and 1.5 %, respectively. Calculate the value of the line-to-ground fault current.

4.2 A Petersen grounding is used in the 66 kV, 50 Hz three-phase transmission lines whose length is found to be 130 km. Each conductor of a transmission line is having a line to earth capacitance whose value is found to be 0.010 μF/km. Calculate the value of the inductance and kVA rating of the Petersen grounding system when a ground fault occurs in a line.

4.3 A 110 kV, 50 Hz, three-phase transmission line passes near the 401 highway from London to Milton in Canada. The length of the transmission line is 140 km and the three conductors are arranged vertically and separated from each other by 1.0 m. Consider that the effective radius of the conductor is 0.001 m. Find the value of the inductance and kVA rating of the Petersen coil when a ground fault occurs in one line.

4.4 A home owner of a village buries a 0.5 m radius hemisphere earth electrode into his neighboring soil whose resistivity is found to be 35 Ω-m. Find the (i) ground resistance of the home and (ii) ground resistance at a distance of 9 m from the center of the hemisphere.

4.5 A sphere electrode is used for the grounding system at a suitable place of a home. The soil resistivity and the ground resistance at that place are found to be 40 Ω-m and 2 Ω, respectively. Calculate the radius of the sphere electrode.

4.6 A 1 in. diameter and 12 ft length cylindrical electrode is driven vertically into the garden soil for the grounding system. The resistivity of the garden soil is found to be 20 Ω-ft. Determine the value of the ground resistance.

4.7 The ground resistance of a 0.5 in. diameter and 12 ft cylindrical electrode for clay soil is found to be 1.5 Ω. Calculate the resistivity of the clay soil using the G.F. Tagg formula.

4.8 The ground resistance of a 1 ft diameter circular plate electrode is found to be 1.5 Ω in silt soil. A person is standing on the soil surface. Calculate the ground resistance of that person.

References

1. H.B. Dwight, Calculation of resistances to ground. AIEE Trans. (Electr. Eng.) **55**, 1319–1328 (1936)
2. G.F. Tagg, *Earth Resistances* (Georges Newnes Limited, London, 1964)
3. A. Gray, *The Theory and Practice of Absolute Measurements in Electricity and Magnetism* (Macmillan and Co., London, 1888)
4. M.A. El-Sharkawi, *Electric Energy* (CRC Press, New York, 2005)

Chapter 5
Soil Resistivity

5.1 Introduction

Soil is defined as the top layer of the earth's crust. It covers most of the land on the earth. It is made up of minerals (rock, sand, clay and silt), air, water and organic materials. The organic materials are formed from the dead plants and animals. There are many properties of soil. Although varying soil properties of different kinds can be observed small area, the electric power utility companies are interested in the electrical properties of the soil, especially the specific resistance or resistivity. Soil resistivity is one of the important factors which plays a vital role in the design and analysis of ground resistance. The dry soil with small particles acts as non-conductor of current. Sands, rocks and loams are some examples of nonconductors. The resistivity of soils drops down when the water content in the soil is more. In this chapter, different types and characteristics of soil, and size of earth electrode and different ground resistance measurement methods will be discussed.

5.2 Soil Resistance and Resistivity

Resistance is the property of a material which opposes the flow of electricity through this material. The mathematical expression of the resistance is,

$$R = \rho \frac{l}{A} \tag{5.1}$$

where l is the length and A is the area and ρ is the specific resistance or resistivity of the conductor. From Eq. (5.1), it is observed that the resistance is mainly depended on the shape and size of the conductor. From Eq. (5.1), the resistivity can be expressed as,

© Springer Science+Business Media Singapore 2016
Md.A. Salam and Q.M. Rahman, *Power Systems Grounding*,
Power Systems, DOI 10.1007/978-981-10-0446-9_5

Fig. 5.1 Simple circuit for
soil resistivity measurement

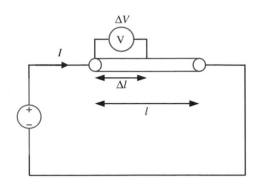

$$\rho = \frac{RA}{l} \qquad (5.2)$$

The unit of the resistivity is ohm-m (Ω-m). From Eq. (5.2), it is seen that the resistivity of a material mainly depends on the area, length and resistance of that material. It is important to study the resistivity of any soil to determine the horizontal and vertical discontinuities in electrical properties of the soil. In practice, the direct current or low frequency alternating current is used to investigate the resistivity of any soil. Figure 5.1 shows a simple circuit for measuring soil resistivity. Here, a voltage source is connected between the two terminals of a conductor or an electrode of length l and area A.

From the measured voltage (ΔV) across a small segment (Δl) of the conductor, the resistivity the soil can be measured as,

$$\rho = \frac{A}{\Delta l} \frac{\Delta V}{I} \qquad (5.3)$$

Based on the types of soil and weather effect the characteristics of the soil vary, and so thus of the soil resistivity. The generalized values of the soil resistivity for different types of soils are shown in Table 5.1.

Table 5.1 Resistivity of
different types of soil

Types of soil	Resistivity (Ω-m)
Sea water	0.15–2
Artesian water	2–12
Clay	2–12
Loams, garden soil	5–50
Clay, sand and gravel mixture	20–250
Concrete	40–1000
Sand	1000–10,000
Moraine gravel	1000–10,000
Ice	10,000–100,000

5.3   Types of Soil

The classification of a soil depends on its smoothness or texture. The texture is the appearance of the soil, and it depends on the relative sizes and shapes of the particles in the soil. According to the grain size, the soils are classified as gravel, sandy, silt, and clay. According to a research finding from MIT, the grain size of the gravel is more than 2 mm. The gravel soil is made up of small pieces of rocks with particles of quartz, feldspar and other minerals. The grain size of the sandy soil is in between 2–0.06 mm and it is also made up of quartz and feldspar. Gravel and sand have very large pores between grains, and as a result, these types of soils can merely retain water and nitrogen molecules. Therefore, both the gravel and the sandy soils are not good choices for the grounding system. The gravel and sandy soils are shown in Figs. 5.2 and 5.3, respectively.

Fig. 5.2 Structure of gravel soil

Fig. 5.3 Structure of sandy soil

Fig. 5.4 Structure of silt soil

The silt soil is made up of fine quartz particles (grains) with a size of 0.06–0.002 mm. The surface area of 1 g silt soil is 15.9 ft^2 which also absorbs and holds less water than the clay soil. Silt has relatively small pores between the grains. Silt is more cohesive and adhesive than the sand. Silt has only limited plasticity and stickiness. The structure of the silt soil is shown in Fig. 5.4.

The clay is one kind of soil that has particles less than 0.002 mm in size and are compact with each other. It is made up of mica, clay minerals and other minerals. It is interesting to note that any soil with grain size in between 0.002 mm and 0.005 m is also referred as clay. The clay has very small pores between the grains. The surface area of 1 g clay soil is 262,467 ft^2 which absorbs and holds sufficient amount of water for a long time. Therefore, this type of soil is good for the grounding system. The clay soil is shown in Fig. 5.5.

Fig. 5.5 Structure of clay soil

5.4 Permeability and Permittivity of Soil

The property of soil that allows water to pass through its pores or voids at the same rate is known as permeability. In sort, the permeability is the ability of the soil to transmit water or air through it. If the permeability of the soil is high, then the water drains out quickly. Whereas, if the permeability of the soil is low, then the water drains out slowly. Different soils have different coefficients of permeability. The coefficients of permeability for different types of soil is shown in Table 5.2.

The permittivity of a soil depends on the frequency, temperature, humidity and composition. The permittivity of a soil is the ability of minerals in the soil to neutralize a part of the static electric field. In this case, different minerals of the soil must contain localized charge that can be displaced by the application of electric field. This displacement is known as polarization. The charge displacement of most of the materials is time dependent, therefore, to explain any system's behaviour, the complex permittivity has to be brought. The electric flux or electric displacement density (D in C/m^2) is directly related to the electric field (E in V/m), and it is written as,

$$D = \varepsilon E \tag{5.4}$$

where, ε is the permittivity, which is expressed as,

$$\varepsilon = \varepsilon_0 \varepsilon_r \tag{5.5}$$

$$\varepsilon_r = \frac{\varepsilon}{\varepsilon_0} \tag{5.6}$$

The relative permeability is the ratio of permeability of medium to the permeability of free space. The relative permeability is equal to the square of the index of refraction. Therefore, the index of refraction (n) can be written as,

$$n^2 = \varepsilon_r = \frac{\varepsilon}{\varepsilon_0} \tag{5.7}$$

$$n = \sqrt{\varepsilon_r} = \sqrt{\frac{\varepsilon}{\varepsilon_0}} \tag{5.8}$$

Table 5.2 Coefficients of permeability

Soil type	Coefficients of permeability (cm/s)
Gravel	1.0–100
Sandy	10^{-3}–10^{-1}
Silt	10^{-8}–10^{-3}
Clay	10^{-10}–10^{-6}

Table 5.3 Permittivity of different soils/parameters

Soil type	Relative permittivity
Air	1
Snow	1.5
Dry loams/clayey soil	2.5
Gravel	5.5
Dry sand	4–6
Wet sand	30
Dry clay	10
Wet clay	27
Organic soil	64
Asphalt	5
Concrete	5.5
Water	80
Dry granite	5
Wet granite	6.5
Frozen soil	6
Frozen sand and gravel	5.5
Forested land	12

where,

D is the electric flux density, C/m^2,
E is the electric field, V/m,
ε is the permittivity of a medium, F/m,
ε_r is the relative permittivity,
n is the index of refraction,
$\varepsilon_0 = 8.854 \times 10^{-12}$ F/m is the permittivity in free space.

The permittivity of different types of soil are shown in Table 5.3.
The complex permittivity can be written as,

$$\varepsilon^* = \varepsilon' + j\varepsilon'' \tag{5.9}$$

where ε' the real is part and ε'' is the imaginary part of the complex permeability.

5.5   Influence of Different Factors on Soil Resistivity

The resistivity of soil mainly depends on the content of water and the resistivity of water itself. Therefore, the resistivity of soil depends on some factors. These are the (i) type of soil along with area, (ii) moisture content, (iii) temperature, (iv) grain size of the material and its distribution, (v) salts dissolved in the soil and (vi) closeness of the minerals in the soil. The moisture of the soil changes in different seasons and at different depths of the soil. The moisture content increases as the depth of the soil increases. The resistivity of the soil increases with a decrease in the moisture content in the soil and vice versa as can be seen in Fig. 5.6. In the heavy winter season, the behavior of the resistivity changes. Therefore, the grounding system needs to be installed at a place which is close to the water level. Lower percentage of salt presence in the soil increases the soil resistivity whereas higher percentage of salt present in the soil reduces the resistivity of the soil as can be seen in Fig. 5.7. However, different types of salt present in the soil have different effects on the resistivity of the soil. The grain size of the soil plays an important role in determining the soil resistivity. The finer grain size of the material along with its distribution reduces soil resistivity.

The temperature versus soil resistivity is shown in Fig. 5.8. When the temperature of the soil reduces to zero or even less, then the water inside the soil becomes frozen. In this case, the conductive cross section of the soil reduces which increases the soil resistivity sharply. The soil resistivity above 0 °C changes slowly. From Fig. 5.8, it is seen that the soil resistivity at negative temperature is much higher than at the positive temperature.

Fig. 5.6 Variation of resistivity with moisture

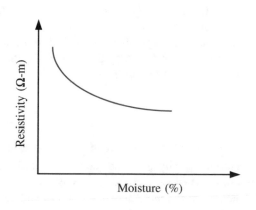

Fig. 5.7 Variation of resistivity with salt

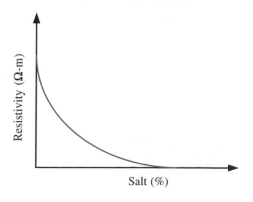

Fig. 5.8 Variation of
resistivity with temperature

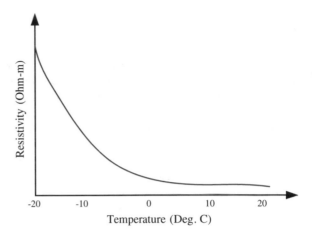

5.6 Current Density of Soil

A constant current field is established in the soil when an ac or dc source injected
through the earth electrode. The existence of current and current density is very
important to identify the behavior of the electric field in the conductor. The current
is defined as the flow of free electrons through a conductor. In an alternative way,
the rate of change of charge (Q) through a given area is known as current. The
current is symbolized by the letter I and its SI unit is ampere (A). Mathematically,
the expression of current is,

$$I = \frac{dQ}{dt} \tag{5.10}$$

In Eq. (5.10), the quantity represents the time unit. The general definition of
current density **J** is the current per unit area (A/m^2). Consider an incremental
current ΔI that is crossing an incremental surface ΔS. This incremental current is
normal to the incremental surface as shown in Fig. 5.9.

The expression of the normal current density can be written as,

$$J_n = \frac{\Delta I}{\Delta S} \tag{5.11}$$

$$\Delta I = J_n \Delta S \tag{5.12}$$

Fig. 5.9 Incremental current
crossing an incremental
surface

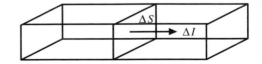

Fig. 5.10 Charge with an
incremental distance

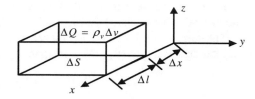

In vector format, Eq. (5.12) can be written as,

$$\Delta I = \mathbf{J} \cdot \Delta \mathbf{S} \tag{5.13}$$

The total current I flowing through the surface S, can be obtained by integrating Eq. (5.13) as,

$$I = \oint_S \mathbf{J} \cdot d\mathbf{S} \tag{5.14}$$

Again, consider that an incremental charge ΔQ is moved to a distance Δx across the yz plane as shown in Fig. 5.10.

The expression of the incremental charge can be written as,

$$\Delta Q = \rho_v \Delta v = \rho_v \Delta S \Delta x \tag{5.15}$$

The expression of the incremental current can be written as,

$$\Delta I = \frac{\Delta Q}{\Delta t} \tag{5.16}$$

Substituting Eq. (5.15) into Eq. (5.16) yields,

$$\Delta I = \frac{\rho_v \Delta S \Delta x}{\Delta t} \tag{5.17}$$

$$\Delta I = \rho_v \Delta S v_x \tag{5.18}$$

where v_x represents the x component of the velocity.

From Eq. (5.18), the current density in the x direction can be expressed as,

$$J_x = \frac{\Delta I}{\Delta S} = \rho_v v_x \tag{5.19}$$

In general, the current density in vector format is,

$$\mathbf{J} = \rho_v \mathbf{v} \tag{5.20}$$

From Eq. (5.20), it is seen that the conventional current density is equal to the product of volume charge density and velocity.

The current density is directly related to the density (ρ_v) of conduction electrons and the drift velocity $(\mathbf{v}_d)$. This relationship can be expressed as,

$$\mathbf{J} = \rho_v \mathbf{v}_d \tag{5.21}$$

The drift velocity is linearly proportional to the electric field vector and it can be expressed as,

$$\mathbf{v}_d = -\mu_e \mathbf{E} \tag{5.22}$$

where μ_e is the mobility of electrons in the given material and its unit is (m^2/Vs).

The current density in terms of concentration of the charge carriers (N_v) and the drift velocity is,

$$\mathbf{J} = N_v(-e)\mathbf{v}_d \tag{5.23}$$

Substituting Eq. (5.22) into Eq. (5.23) yields,

$$\mathbf{J} = N_v(-e)(-\mu_e \mathbf{E}) = N_v e \mu_e \mathbf{E} \tag{5.24}$$

$$\mathbf{J} = \sigma \mathbf{E} \tag{5.25}$$

where σ is the proportionality constant and it is represented as,

$$\sigma = N_v e \mu_e \tag{5.26}$$

Equation (5.25) can be rearranged as,

$$\sigma = \frac{\mathbf{J}}{\mathbf{E}} = \frac{1}{\rho} \tag{5.27}$$

$$\mathbf{E} = \rho \mathbf{J} \tag{5.28}$$

where σ and ρ are the conductivity and resistivity of the soil, respectively. From Eq. (5.28), it is seen that the electric field vector acts along the same direction as the current density vector $\mathbf{J}$. Equation (5.28) can be applied can be applied at any point in the soil.

5.7 Continuity of Earth Current

Figure 5.11 shows a closed surface where a current I_{in} flows into the surface and a current I_{out} flows out of the surface. According to the fundamental definition of current, the expression of the current entering into the closed surface is,

Fig. 5.11 Closed surface
with currents

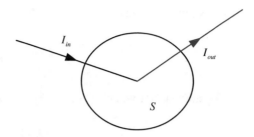

$$I_{in} = \frac{dQ}{dt} \tag{5.29}$$

While, the expression of the current that leaving from the closed surface is,

$$I_{out} = -I_{in} = \frac{-dQ}{dt} \tag{5.30}$$

The total outward current through the closed surface is,

$$I_{out} = \oint_{S} \mathbf{J} \cdot d\mathbf{S} \tag{5.31}$$

Substituting Eq. (5.30) into Eq. (5.31) yields,

$$\oint_{S} \mathbf{J} \cdot d\mathbf{S} = \frac{-dQ}{dt} \tag{5.32}$$

But, the total charge is represented as,

$$Q = \oint_{v} \rho_{v} dv \tag{5.33}$$

Substituting Eq. (5.33) into Eq. (5.34) yields,

$$\oint_{S} \mathbf{J} \cdot d\mathbf{S} = \frac{-d}{dt} \int_{v} \rho_{v} dv \tag{5.34}$$

Applying divergence theorem to change the surface integral into a volume integral as,

$$\oint_{S} \mathbf{J} \cdot d\mathbf{S} = \int_{v} (\nabla \cdot \mathbf{J}) dv \tag{5.35}$$

Substituting Eq. (5.35) into Eq. (5.34) yields,

$$\int_v (\nabla \cdot \mathbf{J}) dv = \frac{-d}{dt} \int_v \rho_v dv \qquad (5.36)$$

For incremental volume, Eq. (5.36) is modified as,

$$(\nabla \cdot \mathbf{J}) \Delta v = \frac{-d\rho_v}{dt} \Delta v \qquad (5.37)$$

$$\nabla \cdot \mathbf{J} = \frac{-d\rho_v}{dt} \qquad (5.38)$$

Equation (5.38) is known as continuity of current equation or simply continuity equation. For steady current, the charge density is constant, and in this case, Eq. (5.38) can be modified as,

$$\nabla \cdot \mathbf{J} = 0 \qquad (5.39)$$

From Eq. (5.39), it is concluded that the total charge leaving the volume is equal to the total charge entering into it. It is also concluded that the current density at any point in the field current is zero. It is meant that the current source is absent in the current field.

Substituting Eq. (5.25) into Eq. (5.38) yields,

$$\nabla \cdot \sigma \mathbf{E} = \frac{-d\rho_v}{dt} \qquad (5.40)$$

Substituting the differential form of Gauss' law ($\nabla \cdot \mathbf{D} = \rho_v$) into Eq. (5.40) yields,

$$\frac{\sigma \rho_v}{\varepsilon} = \frac{-d\rho_v}{dt} \qquad (5.41)$$

$$\frac{\sigma \rho_v}{\varepsilon} + \frac{d\rho_v}{dt} = 0 \qquad (5.42)$$

The general solution of Eq. (5.42) is,

$$\rho_v = k e^{-\left(\frac{\sigma}{\varepsilon}\right)t} \qquad (5.43)$$

Substituting the initial condition (At $t = 0$, $\rho_v = \rho_0$) to Eq. (5.43) yields,

$$\rho_0 = k \qquad (5.44)$$

Again, substituting Eq. (5.44) into Eq. (5.43), the final solution can be written as,

$$\rho_v = \rho_0 e^{-\left(\frac{\sigma}{\varepsilon}\right)t} \tag{5.45}$$

5.8 Current Density at Soil Interface

Figure 5.12 shows the current density vector entering from one region into another region win the soil. At the interface between two regions, this current density vector changes both in magnitude and direction as shown in Fig. 5.12. According to the electric field and electric flux density, some boundary conditions can be derived for the current density vector. From Eq. (5.38), the normal components of current vector can be written as,

$$\mathbf{a}_n \cdot \mathbf{J}_1 - \mathbf{a}_n \cdot \mathbf{J}_2 = \frac{-d\rho_v}{dt} \tag{5.46}$$

For dc (steady) current, $\frac{d\rho_v}{dt} = 0$, then Eq. (5.46) becomes,

$$\mathbf{a}_n \cdot \mathbf{J}_1 - \mathbf{a}_n \cdot \mathbf{J}_2 = 0 \tag{5.47}$$

$$J_{1n} = J_{2n} \tag{5.48}$$

The differential form of the generalized Gauss' law is,

$$\rho_v = \nabla \cdot \mathbf{D} \tag{5.49}$$

$$\rho_v = \nabla \cdot \varepsilon \mathbf{E} \tag{5.50a}$$

Substituting Eq. (5.25) into Eq. (5.50a) yields,

$$\rho = \nabla \cdot \frac{\varepsilon}{\sigma} \mathbf{J} \tag{5.50b}$$

Fig. 5.12 Refraction of steady current lines at the soil interface

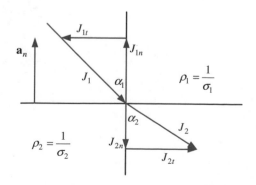

The following rules from the vector identity can be written as:

$$\nabla \cdot (x\mathbf{A}) = \mathbf{A} \cdot \nabla x + x \nabla \cdot \mathbf{A} \tag{5.51}$$

According to Eqs. (5.51), (5.50b) can be expanded as,

$$\rho = \mathbf{J} \cdot \nabla \frac{\varepsilon}{\sigma} + \frac{\varepsilon}{\sigma} \nabla \cdot \mathbf{J} \tag{5.52}$$

Substituting Eq. (5.39) into Eq. (5.52) yields,

$$\rho = \mathbf{J} \cdot \nabla \frac{\varepsilon}{\sigma} \tag{5.53}$$

From Eq. (5.53), the surface charge density in between layers 1 and 2 can be written as,

$$\rho_s = \mathbf{a}_n \cdot \mathbf{D}_1 - \mathbf{a}_n \cdot \mathbf{D}_2 = \frac{\varepsilon_1}{\sigma_1} \mathbf{a}_n \cdot \mathbf{J}_1 - \frac{\varepsilon_2}{\sigma_2} \mathbf{a}_n \cdot \mathbf{J}_2 \tag{5.54}$$

$$\rho_s = \left(\frac{\varepsilon_1}{\sigma_1} - \frac{\varepsilon_2}{\sigma_2} \right) \mathbf{J}_n \tag{5.55}$$

The curl of the electric field is,

$$\nabla \times \mathbf{E} = 0 \tag{5.56}$$

Substituting Eq. (5.25) into Eq. (5.56) yields,

$$\nabla \times \frac{\mathbf{J}}{\sigma} = 0 \tag{5.57}$$

For two mediums, Eq. (5.57) can be written as,

$$\frac{J_{1t}}{\sigma_1} - \frac{J_{2t}}{\sigma_2} = 0 \tag{5.58}$$

$$\frac{J_{1t}}{J_{2t}} = \frac{\sigma_1}{\sigma_2} \tag{5.59}$$

$$\frac{J_{1t}}{J_{2t}} = \frac{\frac{1}{\rho_1}}{\frac{1}{\rho_2}} = \frac{\rho_2}{\rho_1} \tag{5.60}$$

$$J_{1t}\rho_1 = J_{2t}\rho_2 \tag{5.61}$$

From Eq. (5.61), it is concluded that the ratio of tangential components of current density at the interface of two layers is equal to the inverse ratio of their conductivities.

From Fig. 5.14, the following equations can be written as:

$$\cos \alpha_1 = \frac{J_{1n}}{J_1} \tag{5.62}$$

$$\cos \alpha_2 = \frac{J_{2n}}{J_2} \tag{5.63}$$

$$\sin \alpha_1 = \frac{J_{1t}}{J_1} \tag{5.64}$$

$$\sin \alpha_2 = \frac{J_{2t}}{J_2} \tag{5.65}$$

Substituting Eqs. (5.62) and (5.63) into Eq. (5.48) yields,

$$J_1 \cos \alpha_1 = J_2 \cos \alpha_2 \tag{5.66}$$

Substituting Eqs. (5.64) and (5.65) into Eq. (5.59) yields,

$$\frac{J_1 \sin \alpha_1}{J_2 \sin \alpha_2} = \frac{\sigma_1}{\sigma_2} \tag{5.67}$$

$$\sigma_2 J_1 \sin \alpha_1 = \sigma_1 J_2 \sin \alpha_2 \tag{5.68}$$

Dividing Eq. (5.68) by Eq. (5.66) yields,

$$\sigma_2 \tan \alpha_1 = \sigma_1 \tan \alpha_2 \tag{5.69}$$

$$\frac{\tan \alpha_1}{\tan \alpha_2} = \frac{\sigma_1}{\sigma_2} \tag{5.70}$$

The magnitude of current density at the second conductor is,

$$J_2 = \sqrt{J_{2t}^2 + J_{2n}^2} \tag{5.71}$$

Substituting Eqs. (5.64) and (5.65) into Eq. (5.71) yields,

$$J_2 = \sqrt{(J_2 \sin \alpha_2)^2 + (J_2 \cos \alpha_2)^2} \tag{5.72}$$

Substituting Eqs. (5.62) and (5.68) into Eq. (5.72) provides,

$$J_2 = \sqrt{\left(\frac{\sigma_2}{\sigma_1} J_1 \sin \alpha_1\right)^2 + (J_1 \cos \alpha_1)^2} \qquad (5.73)$$

$$J_2 = J_1 \sqrt{\left(\frac{\sigma_2}{\sigma_1} \sin \alpha_1\right)^2 + (\cos \alpha_1)^2} \qquad (5.74)$$

In the presence of the steady current, the tangential electric fields are continuous across the boundary in between two lossy dielectrics and in this condition, it can be written as,

$$E_{1t} = E_{2t} \qquad (5.75)$$

Equation (5.48) can be re-arranged for two layers as,

$$\sigma_1 E_{1n} = \sigma_2 E_{2n} \qquad (5.76)$$

$$E_{1n} = \frac{\sigma_2}{\sigma_1} E_{2n} \qquad (5.77)$$

The surface charge density across two layers is,

$$\rho_s = D_{1n} - D_{2n} = \varepsilon_1 E_{1n} - \varepsilon_2 E_{2n} \qquad (5.78)$$

A surface charge density exists at the interface of the two mediums if the term $\left(\frac{\varepsilon_1}{\sigma_1} - \frac{\varepsilon_2}{\sigma_2}\right)$ of Eq. (5.55) is equal to zero. Then this relationship can be expressed as,

$$\frac{\varepsilon_1}{\sigma_1} - \frac{\varepsilon_2}{\sigma_2} = 0 \qquad (5.79)$$

$$\frac{\sigma_2}{\sigma_1} = \frac{\varepsilon_2}{\varepsilon_1} \qquad (5.80)$$

Substituting Eq. (5.80) into Eq. (5.77) yields,

$$E_{1n} = \frac{\varepsilon_2}{\varepsilon_1} E_{2n} \qquad (5.81)$$

Substituting Eq. (5.81) into Eq. (5.78) yields,

$$\rho_s = \varepsilon_1 \frac{\varepsilon_2}{\varepsilon_1} E_{2n} - \varepsilon_2 E_{2n} \qquad (5.82)$$

Substituting Eq. (5.80) into Eq. (5.82) yields,

$$\rho_s = \left(\varepsilon_1 \frac{\sigma_2}{\sigma_1} - \varepsilon_2 \right) E_{2n} = \left(\varepsilon_1 - \frac{\sigma_1}{\sigma_2} \varepsilon_2 \right) E_{1n} \tag{5.83}$$

If medium 2 is highly conductive than medium 1, then $\sigma_2 > \sigma_1$. So, in this case, $\frac{\sigma_1}{\sigma_2} = 0$. Therefore, Eq. (5.83) becomes,

$$\rho_s = \varepsilon_1 E_{1n} = D_{1n} \tag{5.84}$$

From Eq. (5.85), it is observed that the normal electric flux density of the medium 1 is equal to the surface charge density.

5.9 Derivation of Poisson's and Laplace's Equations

The Poisson's and Laplace's equations are used to calculate the potential and electric field around the high voltage equipment. In this regard, the relationship between the electric field $(\mathbf{E})$ and electrostatic potential (V) is required to derive Poisson's equation and this equation is,

$$\mathbf{E} = -\nabla V \tag{5.85}$$

Taking the divergence of both sides of Eq. (5.85) yields,

$$\nabla . \mathbf{E} = -\nabla . \nabla V \tag{5.86}$$

Substituting the expression of $\mathbf{E} = \frac{\mathbf{D}}{\varepsilon}$ into Eq. (5.86) provides,

$$\nabla . \frac{\mathbf{D}}{\varepsilon} = -\nabla . \nabla V \tag{5.87}$$

$$\frac{1}{\varepsilon} \nabla . \mathbf{D} = -\nabla . \nabla V \tag{5.88}$$

Substituting the differential form of Gauss' law, $\rho_v = \nabla . \mathbf{D}$ into Eq. (5.88) provides,

$$\nabla . \nabla V = -\frac{\rho_v}{\varepsilon} \tag{5.89}$$

According to the rules of vector dot product, the operator $\nabla . \nabla$ can be written as,

$$\nabla.\nabla = \left(\frac{\partial}{\partial x}\mathbf{a}_x + \frac{\partial}{\partial y}\mathbf{a}_y + \frac{\partial}{\partial z}\mathbf{a}_z\right).\left(\frac{\partial}{\partial x}\mathbf{a}_x + \frac{\partial}{\partial y}\mathbf{a}_y + \frac{\partial}{\partial z}\mathbf{a}_z\right) \tag{5.90}$$

$$\nabla.\nabla = \frac{\partial^2}{\partial x^2} + \frac{\partial^2}{\partial y^2} + \frac{\partial^2}{\partial z^2} = \nabla^2 \tag{5.91}$$

Substituting Eq. (5.91) into Eq. (5.89) yields,

$$\nabla^2 V = -\frac{\rho_v}{\varepsilon} \tag{5.92}$$

Equation (5.92) is known as Poisson's equation. If the region contains no free charge, i.e., $\rho_v = 0$, then Eq. (5.92) can be modified as,

$$\nabla^2 V = 0 \tag{5.93}$$

Equation (5.93) is known as Laplace's equation. The Laplace's equation in Cartesian, cylindrical and spherical coordinates can be expressed as,

$$\nabla^2 V = \frac{\partial^2 V}{\partial x^2} + \frac{\partial^2 V}{\partial y^2} + \frac{\partial^2 V}{\partial z^2} = 0 \tag{5.94}$$

$$\nabla^2 V = \frac{1}{\rho}\frac{\partial}{\partial \rho}\left(\rho\frac{\partial V}{\partial \rho}\right) + \frac{1}{\rho^2}\frac{\partial^2 V}{\partial \phi^2} + \frac{\partial^2 V}{\partial z^2} = 0 \tag{5.95}$$

$$\nabla^2 V = \frac{1}{r^2}\frac{\partial}{\partial r}\left(r^2\frac{\partial V}{\partial r}\right) + \frac{1}{r^2 \sin\theta}\frac{\partial}{\partial\theta}\left(\sin\theta\frac{\partial V}{\partial\theta}\right) + \frac{1}{r^2 \sin^2\theta}\frac{\partial^2 V}{\partial \phi^2} = 0 \tag{5.96}$$

Example 5.1 The expression of electric potential in Cartesian coordinates is given by $V(x, y, z) = 2x^2 y + 3z^2$. Calculate the (i) numerical value of the voltage at point $P(1, 3, 2)$, (ii) electric field, and (iii) verify the Laplace equation.

Solution

(i) The numerical value of the potential can be determined as,

$$V(x, y, z) = 2(1)^2 3 + 3(2)^2 = 18 \text{ V}$$

(ii) The expression of electric field is calculated as,

$$\mathbf{E} = -\left(\frac{\partial V}{\partial x}\mathbf{a}_x + \frac{\partial V}{\partial y}\mathbf{a}_y + \frac{\partial V}{\partial z}\mathbf{a}_z\right)$$

$$\frac{\partial V}{\partial x} = \frac{\partial (2x^2 y + 3z^2)}{\partial x} = 4xy$$

$$\frac{\partial V}{\partial y} = \frac{\partial (2x^2 y + 3z^2)}{\partial y} = 2x^2$$

$$\frac{\partial V}{\partial z} = \frac{\partial (2x^2 y + 3z^2)}{\partial z} = 6z$$

$$\mathbf{E} = -\left(4xy\mathbf{a}_x + 2x^2\mathbf{a}_y + 6z\mathbf{a}_z\right) \text{ V/m}$$

The electric field at point $P(1, 3, 2)$ is,

$$\mathbf{E} = -\left(12\mathbf{a}_x + 2\mathbf{a}_y + 12\mathbf{a}_z\right) \text{ V/m}$$

(iii) The derivatives of the respective variables are,

$$\frac{\partial V}{\partial x} = \frac{\partial}{\partial x}(2x^2 y + 3z^2) = 4xy$$

$$\frac{\partial^2 V}{\partial x^2} = \frac{\partial}{\partial x}(4xy) = 4y$$

$$\frac{\partial V}{\partial y} = \frac{\partial}{\partial y}(2x^2 y + 3z^2) = 2x^2$$

$$\frac{\partial^2 V}{\partial y^2} = \frac{\partial}{\partial y}(2x^2) = 0$$

$$\frac{\partial V}{\partial z} = \frac{\partial}{\partial z}(2x^2 y + 3z^2) = 6z$$

$$\frac{\partial^2 V}{\partial z^2} = \frac{\partial}{\partial z}(6z) = 6$$

$$\nabla^2 V = \frac{\partial^2 V}{\partial x^2} + \frac{\partial^2 V}{\partial y^2} + \frac{\partial^2 V}{\partial z^2} = 4y + 6 = 4 \times 3 + 6 = 18$$

Thus, it does not satisfy Laplace's equation.

Practice problem 5.1

The expression of an electric potential in cylindrical coordinates is given by $V(\rho, \phi, z) = 3\rho^2 z \sin \phi$. Find the (i) numerical value of the voltage at point $P(\rho = 1, \phi = 30°, z = 3)$, (ii) electric field at point $P(\rho = 1, \phi = 30°, z = 3)$, and (iii) verify the Laplace equation.

5.10 Uniqueness Theorem

Each electrostatic object has its own boundary, and this boundary is known as boundary potentials. The solution of a quadratic equation must be unique if it satisfies its related boundary conditions. Therefore, any solution of Laplace's equation that satisfies the boundary conditions is known as the uniqueness theorem.

Consider a finite volume v is bounded by a closed surface s as shown in Fig. 5.13. To prove the uniqueness theorem, we assume that there are two solutions of Laplace's equation. These solutions are,

$$\nabla^2 V_1 = 0 \tag{5.97}$$

$$\nabla^2 V_2 = 0 \tag{5.98}$$

Subtracting Eq. (5.98) from Eq. (5.97) yields,

$$\nabla^2 (V_1 - V_2) = 0 \tag{5.99}$$

The potential at the boundary of the surface must be identical and it can be expressed as,

$$V|_b = V_1|_b = V_2|_b \tag{5.100}$$

The following vector identity is used to verify the uniqueness theorem:

$$f.\mathbf{A} = f(\nabla.\mathbf{A}) + \mathbf{A}.(\nabla f) \tag{5.101}$$

In this case, consider that f is a scalar function and $\mathbf{A}$ is a vector function. The following functions can be defined as:

$$f = V_1 - V_2 \tag{5.102}$$

$$\mathbf{A} = \nabla(V_1 - V_2) \tag{5.103}$$

Fig. 5.13 Surface with a small volume

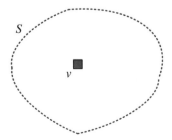

Substituting Eqs. (5.102) and (5.103) into Eq. (5.101) yields,

$$(V_1 - V_2).\nabla(V_1 - V_2) = (V_1 - V_2)[\nabla.\nabla(V_1 - V_2)] + \nabla(V_1 - V_2).\nabla(V_1 - V_2)$$
$$(5.104)$$

$$(V_1 - V_2).\nabla(V_1 - V_2) = (V_1 - V_2)[\nabla^2(V_1 - V_2)] + [\nabla(V_1 - V_2)]^2 \quad (5.105)$$

Integrating Eq. (5.105) over the volume v yields,

$$\int_v (V_1 - V_2).\nabla(V_1 - V_2)dv = \int_v (V_1 - V_2)[\nabla^2(V_1 - V_2)]dv + \int_v [\nabla(V_1 - V_2)]^2 dv$$
$$(5.106)$$

Applying the divergence theorem to replace the volume integral of the left side of the Eq. (5.106), it becomes,

$$\int_v (V_1 - V_2).\nabla(V_1 - V_2)dv = \oint_s [(V_1 - V_2)]_b[\nabla(V_1 - V_2)]_b.\, dS \quad (5.107)$$

Substituting Eq. (5.107) into Eq. (5.106) yields,

$$\oint_s [(V_1 - V_2)]_b[\nabla(V_1 - V_2)]_b.\, dS = \int_v (V_1 - V_2)[\nabla^2(V_1 - V_2)]dv + \int_v [\nabla(V_1 - V_2)]^2 dv$$
$$(5.108)$$

By hypothesis, the first and second integrals of Eq. (5.108) are equal to zero, and in this case, Eq. (5.108) can be modified as,

$$\int_v [\nabla(V_1 - V_2)]^2 dv = 0 \quad\quad\quad\quad (5.109)$$

$$\nabla(V_1 - V_2) = 0 \quad\quad\quad\quad (5.110)$$

If the gradient of $(V_1 - V_2)$ is zero everywhere in the closed surface then $(V_1 - V_2)$ does not change with any coordinates. In this case, Eq. (5.110) can be represented as,

$$V_1 - V_2 = \text{constant} \quad\quad\quad\quad (5.111)$$

The constant term in Eq. (5.111) can be determined by considering a specific point on the boundary of the object. If the constant term is zero at that specific point, then Eq. (5.111) becomes,

$$V_1 = V_2 \tag{5.112}$$

Equation (5.112) generally provides two identical solutions. The uniqueness theorem can also be applied to Poisson's equation as,

$$\nabla^2 V_1 = -\frac{\rho_v}{\varepsilon} \tag{5.113}$$

$$\nabla^2 V_2 = -\frac{\rho_v}{\varepsilon} \tag{5.114}$$

Subtracting Eq. (5.114) into Eq. (5.113) yields,

$$\nabla^2 (V_1 - V_1) = -\frac{\rho_v}{\varepsilon} + \frac{\rho_v}{\varepsilon} = 0 \tag{5.115}$$

The solution of Eq. (5.115) can be obtained by considering proper boundary conditions.

5.11 Solutions of Laplace's Equation

Direct integration and differentiation methods are used to solve Laplace's equation. The solutions of Laplace's equation for one-dimension, two-dimension, and three-dimension are discussed below in details:

5.11.1 One Dimension Solution

In one-dimension solution, let us consider the potential, V varies only in the x-direction. Then, the Laplace equation can be modified as,

$$\frac{\partial^2 V}{\partial x^2} = 0 \tag{5.116}$$

The partial derivative of Eq. (5.116) can be represented by an ordinary differential equation and it can be written as,

$$\frac{d^2 V}{dx^2} = 0 \tag{5.117}$$

Integrating Eq. (5.117) yields,

$$\frac{dV}{dx} = C \tag{5.118}$$

Again, integrating Eq. (5.118) yields,

$$\int dV = C \int dx \tag{5.119}$$

$$V = Cx + D \tag{5.120}$$

Equation (5.120) is the solution of Laplace equation in the x-direction and C and D are the integrating constants and these can be determined by using the appropriate boundary conditions.

Again, consider the potential V in cylindrical coordinates which varies only in the ρ-direction. Then, the Laplace equation can be written as,

$$\frac{1}{\rho}\frac{\partial}{\partial \rho}\left(\rho \frac{\partial V}{\partial \rho}\right) = 0 \tag{5.121}$$

The partial derivative of Eq. (5.121) can be represented by an ordinary differential equation, and it can be expressed as,

$$\frac{1}{\rho}\frac{d}{d\rho}\left(\rho \frac{dV}{d\rho}\right) = 0 \tag{5.122}$$

$$\frac{d}{d\rho}\left(\rho \frac{dV}{d\rho}\right) = 0 \tag{5.123}$$

Integrating Eq. (5.123) yields,

$$\rho \frac{dV}{d\rho} = A \tag{5.124}$$

$$dV = A \frac{d\rho}{\rho} \tag{5.125}$$

Integrating Eq. (5.125) yields,

$$V = A \ln \rho + B \tag{5.126}$$

Equation (5.126) is the solution of Laplace equation in the ρ-direction.

Consider that the potential V in spherical coordinates varies only in the r-direction. Then the Laplace equation can be written as,

$$\frac{\partial}{\partial r}\left(r^2\frac{\partial V}{\partial r}\right) = 0 \tag{5.127}$$

The partial derivative of Eq. (5.127) can be represented by an ordinary differential equation and it can be expressed as,

$$\frac{d}{dr}\left(r^2\frac{dV}{dr}\right) = 0 \tag{5.128}$$

Integrating Eq. (5.128) yields,

$$r^2\frac{dV}{dr} = k_1 \tag{5.129}$$

$$dV = k_1 r^{-2} dr \tag{5.130}$$

Again, integrating Eq. (5.130) yields,

$$V = k_2 - k_1\frac{1}{r} \tag{5.131}$$

Equation (5.131) is the solution of Laplace equation in the r-direction.

5.11.2   Two-Dimension Solution

In two-dimension solution, let us consider that the potential V varies only in the x and y directions. Then, the Laplace equation in rectangular form can be written as,

$$\frac{\partial^2 V}{\partial x^2} + \frac{\partial^2 V}{\partial y^2} = 0 \tag{5.132}$$

The partial derivative of Eq. (5.132) can be represented by an ordinary differential equation and it can be expressed as,

$$\frac{d^2 V}{dx^2} + \frac{d^2 V}{dy^2} = 0 \tag{5.133}$$

Consider that the general solution of Eq. (5.133) is,

$$V(x, y) = X(x)Y(y) \tag{5.134}$$

Substituting Eq. (5.134) into Eq. (5.133) yields,

$$Y\frac{d^2X}{dx^2} + X\frac{d^2Y}{dy^2} = 0 \tag{5.135}$$

Dividing Eq. (5.135) by XY yields,

$$\frac{1}{X}\frac{d^2X}{dx^2} + \frac{1}{Y}\frac{d^2Y}{dy^2} = 0 \tag{5.136}$$

It is seen that the first part of Eq. (5.136) is a function of x, and it is equal to a constant term. Similarly, the second part is a function of y, and it is equal to another constant term. The following equations can be written as:

$$\frac{1}{X}\frac{d^2X}{dx^2} = A_1^2 \tag{5.137}$$

$$\frac{1}{Y}\frac{d^2Y}{dy^2} = B_1^2 \tag{5.138}$$

Equation (5.136) is then modified to,

$$A_1^2 + B_1^2 = 0 \tag{5.139}$$

$$A_1^2 = -B_1^2 \tag{5.140}$$

Equation (5.137) can be re-arranged as,

$$\frac{d^2X}{dx^2} - XA_1^2 = 0 \tag{5.141}$$

Consider $\frac{d}{dx} = D$, then Eq. (5.141) can be modified to,

$$D^2 - A_1^2 = 0 \tag{5.142}$$

$$D = \pm A_1 \tag{5.143}$$

$$DX = \pm A_1X \tag{5.144}$$

$$\frac{dX}{dx} = A_1X \tag{5.145}$$

$$\frac{dX}{X} = A_1dx \tag{5.146}$$

Integrating Eq. (5.146) yields,

$$\ln X = A_1 x + k \tag{5.147}$$

$$X = e^k e^{A_1 x} = k_3 e^{A_1 x} \tag{5.148}$$

where $k_3 = e^k$

Similarly, the other solution is,

$$X = k_4 e^{-A_1 x} \tag{5.149}$$

In general, the solution is,

$$X(x) = k_3 e^{A_1 x} + k_4 e^{-A_1 x} \tag{5.150}$$

Since, $\cosh A_1 x = \frac{e^{A_1 x} + e^{-A_1 x}}{2}$ and $\sinh A_1 x = \frac{e^{A_1 x} - e^{-A_1 x}}{2}$, the following equations can be written as:

$$e^{A_1 x} = \cosh A_1 x + \sinh A_1 x \tag{5.151}$$

$$e^{-A_1 x} = \cosh A_1 x - \sinh A_1 x \tag{5.152}$$

The solution of Eq. (5.141) is,

$$X(x) = k_1 \cosh A_1 x + k_2 \sinh A_1 x \tag{5.153}$$

where $k_1 = k_3 + k_4$ and $k_2 = k_3 - k_4$.

Equation (5.138) can be rearranged as,

$$\frac{d^2 Y}{dy^2} = B_1{}^2 Y \tag{5.154}$$

Substituting Eq. (5.140) into Eq. (5.154) yields,

$$\frac{d^2 Y}{dy^2} = -A_1{}^2 Y \tag{5.155}$$

$$D^2 = -A_1{}^2 \tag{5.156}$$

$$D = \pm j A_1 \tag{5.157}$$

Therefore, the solution of Eq. (5.155) becomes,

$$Y(y) = c_3 e^{jA_1 y} + c_4 e^{-jA_1 y} \qquad (5.158)$$

Since, $\cos A_1 y = \frac{e^{jA_1 y} + e^{-jA_1 y}}{2}$ and $\sin A_1 y = \frac{e^{jA_1 y} - e^{-jA_1 y}}{j2}$, the following equations can be written as:

$$e^{jA_1 y} = \cos A_1 y + j \sin A_1 y \qquad (5.159)$$

$$e^{-jA_1 y} = \cos A_1 y - j \sin A_1 y \qquad (5.160)$$

Equation (5.158) can be modified as,

$$Y(y) = k_5 \cos A_1 y + k_6 \sin A_1 y \qquad (5.161)$$

where $k_5 = c_3 + c_4$ band $k_6 = c_3 - jc_4$.

The two-dimension solution of Laplace equation is,

$$V(x, y) = (k_1 \cosh A_1 x + k_2 \sinh A_1 x)(k_5 \cos A_1 y + k_6 \sin A_1 y) \qquad (5.162)$$

Consider Fig. 5.14 to determine the constants k_1, k_2, k_5 and k_6. The boundary conditions of Fig. 5.14 are,

$$V = 0 \text{ at } x = 0$$
$$V = V_0 \text{ at } x = a$$
$$V = 0 \text{ at } y = 0$$
$$V = 0 \text{ at } y = b$$

Apply the third boundary condition ($V = 0$ at $y = 0$) to Eq. (5.163) yields,

$$0 = (k_1 \cosh A_1 x + k_2 \sinh A_1 x)(k_5 + 0) \qquad (5.163)$$

$$k_5 = 0 \qquad (5.164)$$

Fig. 5.14 A rectangular conducting object

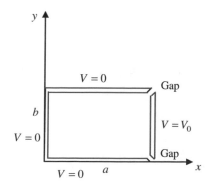

Substituting Eq. (5.164) into Eq. (5.162) yields,

$$V(x, y) = (k_1 \cosh A_1 x + k_2 \sinh A_1 x)(k_6 \sin A_1 y) \tag{5.165}$$

Applying the fourth boundary condition ($V = 0$ at $y = b$) to Eq. (5.162) yields,

$$0 = \sin A_1 b \tag{5.166}$$

$$\sin m\pi = \sin A_1 b \tag{5.167}$$

where $m = 0, 1, 2, \ldots$

$$A_1 = \frac{m\pi}{b} \tag{5.168}$$

Substituting Eq. (5.168) into Eq. (5.165) yields,

$$V(x, y) = \left(k_1 \cosh \frac{m\pi}{b} x + k_2 \sinh \frac{m\pi}{b} x\right)\left(k_6 \sin \frac{m\pi}{b} y\right) \tag{5.169}$$

Applying the first boundary condition ($V = 0$ at $x = 0$) to Eq. (5.169) yields,

$$0 = (k_1 + 0)\left(k_6 \sin \frac{m\pi}{b} y\right) \tag{5.170}$$

$$k_1 = 0 \tag{5.171}$$

Substituting Eq. (5.171) into Eq. (5.169) yields,

$$V(x, y) = k_2 k_6 \sinh\left(\frac{m\pi}{b} x\right) \sin\left(\frac{m\pi}{b} y\right) \tag{5.172}$$

$$V(x, y) = k \sinh\left(\frac{m\pi}{b} x\right) \sin\left(\frac{m\pi}{b} y\right) \tag{5.173}$$

where $k = k_2 k_6$

Again, applying the boundary condition $V = V_0$ at $x = a$ into Eq. (5.173) yields,

$$V(a, y) = V_0 = k \sinh\left(\frac{m\pi}{b} a\right) \sin\left(\frac{m\pi}{b} y\right) \tag{5.174}$$

For an infinite series, Eq. (5.173) can be written as,

$$V(x, y) = \sum_{m=1}^{\infty} k \sinh\left(\frac{m\pi}{b} x\right) \sin\left(\frac{m\pi}{b} y\right) \tag{5.175}$$

$$V_0 = \sum_{m=1}^{\infty} k \sinh\left(\frac{m\pi}{b}a\right) \sin\left(\frac{m\pi}{b}y\right) \tag{5.176}$$

Multiplying both sides by $\sin\left(\frac{n\pi y}{b}\right)$ of Eq. (5.176) and integrating over $0<y<b$ yields,

$$\int_0^b V_0 \sin\left(\frac{n\pi}{b}y\right)dy = \sum_{m=1}^{\infty} k \sinh\left(\frac{m\pi}{b}a\right)\int_0^b \sin\left(\frac{m\pi}{b}y\right)\sin\left(\frac{n\pi}{b}y\right)dy \tag{5.177}$$

The orthogonal product rule is,

$$\int_0^b \sin(my)\sin(ny)dy = \begin{vmatrix} 0 & m \neq n \\ \frac{1}{2} & m = n \end{vmatrix} \tag{5.178}$$

Applying the rules of Eq. (5.178) into Eq. (5.177) yields,

$$\int_0^b V_0 \sin\left(\frac{m\pi}{b}y\right)dy = \sum_{m=1}^{\infty} k \sinh\left(\frac{m\pi}{b}a\right)\int_0^b \sin^2\left(\frac{m\pi}{b}y\right)dy \tag{5.179}$$

$$\int_0^b V_0 \sin\left(\frac{m\pi}{b}y\right)dy = k \sinh\left(\frac{m\pi}{b}a\right)\frac{1}{2}\int_0^b \left(1 - \cos\frac{m\pi}{b}y\right)dy \tag{5.180}$$

$$-V_0 \frac{b}{m\pi}\left[\cos\left(\frac{m\pi}{b}y\right)\right]_0^b = k\frac{b}{2}\sinh\left(\frac{m\pi}{b}a\right) \tag{5.181}$$

$$k \sinh\left(\frac{m\pi}{b}a\right) = \frac{2V_0}{m\pi}[1 - \cos m\pi] \tag{5.182}$$

$$k \sinh\left(\frac{m\pi}{b}a\right) = \begin{vmatrix} \frac{4V_0}{m\pi} & m = 1, 3, 5,\ldots \\ 0 & m = 2, 4, 6,\ldots \end{vmatrix} \tag{5.183}$$

$$k = \frac{4V_0}{m\pi \sinh\left(\frac{m\pi}{b}a\right)} \quad \text{for } m = \text{odd} \tag{5.184}$$

$$k = 0 \quad \text{for } m = \text{even} \tag{5.185}$$

Substituting Eq. (5.184) into Eq. (5.175) gives the complete solution as,

$$V(x, y) = \frac{4V_0}{\pi} \sum_{m=1,3,5,\ldots}^{\infty} \frac{\sinh\left(\frac{m\pi}{b}x\right)\sin\left(\frac{m\pi}{b}y\right)}{m \sinh\left(\frac{m\pi}{b}a\right)} \qquad (5.186)$$

Example 5.2 The boundary conditions of regions 1 and 2 are defined as $y = 0$, $V = 0$ and $y = l$, $V = V_0$, respectively. Use Laplace equation to determine the expression of voltage.

Solution

For region 1, the Laplace equation is,

$$\frac{\partial^2 V_1}{\partial y^2} = 0$$

The solution is,

$$V_1 = Ay + B$$

At $y = 0$, $V = 0$, then $B = 0$
For region 2, the Laplace equation is,

$$\frac{\partial^2 V_2}{\partial y^2} = 0$$

The solution is,

$$V_2 = Ay + B$$

At $y = l$, $V = V_0$, then $A = \frac{V_0}{l}$
Let $V_1 = V_2 = V$, then the expression of voltage is,

$$V = \frac{V_0 y}{l}$$

Example 5.3 Calculate the potential of a rectangular object of infinite length. Consider the conditions $a = b = 1$ m, $V_0 = 100$ V, $x = \frac{a}{2}$ and $y = \frac{b}{2}$. Also, find the electric field intensity.

Solution

The voltage can be determined as,

$$V\left(\frac{a}{2},\frac{b}{2}\right) = \frac{400}{\pi}\left[\frac{\sinh\left(\frac{\pi}{2}\right)\sin\left(\frac{\pi}{2}\right)}{\sinh(\pi)} + \frac{\sinh\left(\frac{3\pi}{2}\right)\sin\left(\frac{3\pi}{2}\right)}{3\sinh(3\pi)} + \frac{\sinh\left(\frac{5\pi}{2}\right)\sin\left(\frac{5\pi}{2}\right)}{5\sinh(5\pi)}\right]$$

$$V\left(\frac{a}{2},\frac{b}{2}\right) = \frac{400}{\pi}[0.1992 - 0.00299 + 0.0000776] = 25\text{ V}$$

The electric field intensity can be determined as,

$$\mathbf{E} = -\nabla V = -\frac{\partial V}{\partial x}\mathbf{a}_x - \frac{\partial V}{\partial y}\mathbf{a}_y$$

$$\mathbf{E} = -\frac{400}{b}\left[\left\{\frac{\cosh\left(\frac{m\pi}{b}x\right)\sin\left(\frac{m\pi}{b}y\right)}{\sinh\left(\frac{m\pi}{b}a\right)}\mathbf{a}_x\right\} + \left\{\frac{\sinh\left(\frac{m\pi}{b}x\right)\cos\left(\frac{m\pi}{b}y\right)}{\sinh\left(\frac{m\pi}{b}a\right)}\mathbf{a}_y\right\}\right]$$

$$\mathbf{E} = -\frac{400}{b}[(0.217 - 0.0089 + 0.00038)\mathbf{a}_x + 0\mathbf{a}_y]$$

$$\mathbf{E} = -83.39\mathbf{a}_x\text{ V/m}$$

Practice problem 5.2
The potential is a function of ϕ in cylindrical coordinates of the radial planes. The boundary conditions of these planes are given as $V = 0$ at $\phi = 0$ and $V = V_0$ at $\phi = \gamma$. Determine the potential and electric field intensity.

Practice problem 5.3
Determine the potential of a rectangular object of infinite length. Consider the boundary conditions $a = 2b = 4\text{ m}$, $V_0 = 100\text{ V}$, $x = a$ and $y = \frac{b}{2}$.

5.12   Solution of Laplace's Equation in Cylindrical Coordinates

Since most of the earth electrodes look like a cylinder, attention is given in finding the solution of Laplace equation in cylindrical coordinates to calculate the potential and electric field distribution around any earth electrode. In cylindrical coordinates, Laplace's equation for electrostatic potential V is given as,

$$\nabla^2 V = \frac{1}{\rho}\frac{\partial}{\partial\rho}\left(\rho\frac{\partial V}{\partial\rho}\right) + \frac{1}{\rho^2}\frac{\partial^2 V}{\partial\phi^2} + \frac{\partial^2 V}{\partial z^2} = 0 \qquad (5.187)$$

The general solution of Eq. (5.187) is,

$$V(\rho, \phi, z) = R(\rho)\Phi(\phi)Z(z) \tag{5.188}$$

Substituting Eq. (5.188) into Eq. (5.187) yields,

$$\Phi Z \frac{1}{\rho} \frac{d}{d\rho} \left(\rho \frac{dR}{d\rho} \right) + RZ \frac{1}{\rho^2} \frac{d^2\Phi}{d\phi^2} + R\Phi \frac{d^2Z}{dz^2} = 0 \tag{5.189}$$

Dividing Eq. (5.189) by the term $R\Phi Z$ provides,

$$\frac{1}{R\rho} \frac{d}{d\rho} \left(\rho \frac{dR}{d\rho} \right) + \frac{1}{\Phi\rho^2} \frac{d^2\Phi}{d\phi^2} + \frac{1}{Z} \frac{d^2Z}{dz^2} = 0 \tag{5.190}$$

$$\frac{1}{R\rho} \frac{d}{d\rho} \left(\rho \frac{dR}{d\rho} \right) + \frac{1}{\Phi\rho^2} \frac{d^2\Phi}{d\phi^2} = -\frac{1}{Z} \frac{d^2Z}{dz^2} \tag{5.191}$$

The right side term of Eq. (5.191) is only a function of z, and it can be defined as,

$$-\frac{1}{Z} \frac{d^2Z}{dz^2} = -k^2 \tag{5.192}$$

$$\frac{d^2Z}{dz^2} + zk^2 = 0 \tag{5.193}$$

The solution of Eq. (5.193) is,

$$Z(z) = Ae^{kz} + Be^{-kz} \tag{5.194}$$

$$Z(z) = A_z \cosh(kz) + B_z \sinh(kz) \tag{5.195}$$

Substituting Eq. (5.192) into Eq. (5.191) yields,

$$\frac{1}{R\rho} \frac{d}{d\rho} \left(\rho \frac{dR}{d\rho} \right) + \frac{1}{\Phi\rho^2} \frac{d^2\Phi}{d\phi^2} = -k^2 \tag{5.196}$$

$$\frac{\rho}{R} \frac{d}{d\rho} \left(\rho \frac{dR}{d\rho} \right) + \rho^2 k^2 = -\frac{1}{\Phi} \frac{d^2\Phi}{d\phi^2} \tag{5.197}$$

Again, the right side of Eq. (5.197) is a function ϕ and it is represented by m^2. Then, the following equation can be written as:

$$-\frac{1}{\Phi} \frac{d^2\Phi}{d\phi^2} = m^2 \tag{5.198}$$

$$\frac{d^2\Phi}{d\phi^2} + \Phi m^2 = 0 \tag{5.199}$$

The solution of Eq. (5.199) can be written as,

$$\Phi(\phi) = Ae^{jm\phi} + Be^{-jm\phi} \tag{5.200}$$

$$\Phi(\phi) = A_\phi \cos(m\phi) + B_\phi \sin(m\phi) \tag{5.201}$$

Equation (5.197) can be modified as,

$$\frac{\rho}{R}\frac{d}{d\rho}\left(\rho\frac{dR}{d\rho}\right) + \rho^2 k^2 - m^2 = 0 \tag{5.202}$$

$$\rho\frac{d}{d\rho}\left(\rho\frac{dR}{d\rho}\right) + (\rho^2 k^2 - m^2)R = 0 \tag{5.203}$$

$$\rho^2\frac{d^2R}{d\rho^2} + \rho\frac{dR}{d\rho} + (\rho^2 k^2 - m^2)R = 0 \tag{5.204}$$

$$\frac{d^2R}{d\rho^2} + \frac{1}{\rho}\frac{dR}{d\rho} + \left(k^2 - \frac{m^2}{\rho^2}\right)R = 0 \tag{5.205}$$

The final solution of Eq. (5.205) is,

$$R = B_1 J_n(m\rho) + B_2 N_n(m\rho) \tag{5.206}$$

where,
$J_n(m\rho)$ is the Bessel function of the first kind of order n with argument $m\rho$,
$N_n(m\rho)$ is the Bessel function of the second kind of order n with argument $m\rho$.

5.13 Spherical Coordinate System

Any point in a spherical coordinates is defined as $M(r, \theta, \phi)$. A radial line with the length of r is drawn at an angle θ with the z-axis and the unit vectors of this system are $\mathbf{a}_r$, $\mathbf{a}_\theta$ and $\mathbf{a}_\phi$ as shown in Fig. 2.23a. Here, $\mathbf{a}_r$ is parallel to the radial line and the unit vector $\mathbf{a}_\phi$ is tangent to the sphere which increases along the direction of increasing ϕ. The unit vector $\mathbf{a}_\theta$ is basically a tangent to the sphere which is not shown in Fig. 5.15 and it increases along the direction of increasing θ. The vector $\mathbf{A}$ in terms of spherical components can be written as [1–4],

Fig. 5.15 Spherical
coordinates with unit vectors

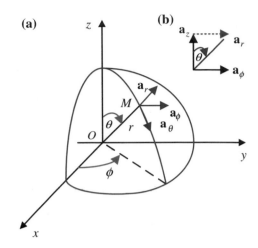

$$\mathbf{A} = A_r\mathbf{a}_r + A_\theta\mathbf{a}_\theta + A_\phi\mathbf{a}_\phi \equiv (A_r, A_\theta, A_\phi) \tag{5.207}$$

The ranges of the coordinates are,

$$0 < r < \infty, \quad 0 < \theta < \pi, \quad 0 < \phi < 2\pi \tag{5.208}$$

The magnitude of the vector can be written as,

$$|\mathbf{A}| = \sqrt{A_r^2 + A_\theta^2 + A_\phi^2} \tag{5.209}$$

In this coordinate system, the properties of unit vectors are,

$$\mathbf{a}_r \cdot \mathbf{a}_r = \mathbf{a}_\theta \cdot \mathbf{a}_\theta = \mathbf{a}_\phi \cdot \mathbf{a}_\phi = 1 \tag{5.210}$$

$$\mathbf{a}_r \cdot \mathbf{a}_\theta = \mathbf{a}_\theta \cdot \mathbf{a}_\phi = \mathbf{a}_\phi \cdot \mathbf{a}_r = 0 \tag{5.211}$$

From Fig. 5.16, the following expressions can be written as:

$$\rho = r \sin\theta \tag{5.212}$$

$$x = \rho \cos\phi \tag{5.213}$$

$$y = \rho \sin\phi \tag{5.214}$$

$$z = r \cos\theta \tag{5.215}$$

Fig. 5.16 Relationship
between Cartesian and
spherical coordinates

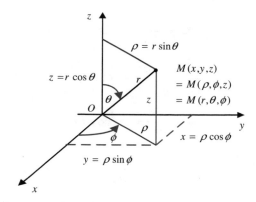

Substituting Eq. (5.212) into Eqs. (5.213) and (5.214) yields,

$$x = r \sin \theta \cos \phi \tag{5.216}$$

$$y = r \sin \theta \sin \phi \tag{5.217}$$

Again, consider Fig. 5.16 to derive the relationship between the Cartesian and spherical coordinates.

From Eqs. (5.215) to (5.217), the following relationships can be written as:

$$r = \sqrt{x^2 + y^2 + z^2} \tag{5.218}$$

$$\theta = \cos^{-1}\left(\frac{z}{\sqrt{x^2 + y^2 + z^2}}\right) \tag{5.219}$$

$$\phi = \tan^{-1}\left(\frac{y}{x}\right) \tag{5.220}$$

Again, consider that the position vector in Cartesian coordinates is,

$$\mathbf{r} = x\mathbf{a}_x + y\mathbf{a}_y + z\mathbf{a}_z \tag{5.221}$$

Substituting Eqs. (5.215), (5.216) and (5.217) into Eq. (5.221) yields,

$$\mathbf{r} = r \sin \theta \cos \phi \mathbf{a}_x + r \sin \theta \sin \phi \mathbf{a}_y + r \cos \theta \mathbf{a}_z \tag{5.222}$$

Differentiating Eq. (5.222) with respect to the r, θ and ϕ yields,

$$\frac{\partial \mathbf{r}}{\partial r} = \sin \theta \cos \phi \mathbf{a}_x + \sin \theta \sin \phi \mathbf{a}_y + \cos \theta \mathbf{a}_z \tag{5.223}$$

$$n_r = \left| \frac{\partial \mathbf{r}}{\partial r} \right| = \sqrt{\sin^2 \theta \cos^2 \phi + \sin^2 \theta \sin^2 \phi + \cos^2 \theta} = 1 \qquad (5.224)$$

$$\frac{\partial \mathbf{r}}{\partial \theta} = r \cos \theta \cos \phi \mathbf{a}_x + r \cos \theta \sin \phi \mathbf{a}_y - r \sin \theta \mathbf{a}_z \qquad (5.225)$$

$$n_\theta = \left| \frac{\partial \mathbf{r}}{\partial \theta} \right| = r \sqrt{\cos^2 \theta \cos^2 \phi + \cos^2 \theta \sin^2 \phi + \sin^2 \theta} = r \qquad (5.226)$$

$$\frac{\partial \mathbf{r}}{\partial \phi} = -r \sin \theta \sin \phi \mathbf{a}_x + r \sin \theta \cos \phi \mathbf{a}_y \qquad (5.227)$$

$$n_\phi = \left| \frac{\partial \mathbf{r}}{\partial \phi} \right| = r \sqrt{\sin^2 \theta \sin^2 \phi + \sin^2 \theta \cos^2 \phi} = r \sin \theta \qquad (5.228)$$

The unit vector $\mathbf{a}_r$ is defined as,

$$\hat{r} = \mathbf{a}_r = \frac{\frac{\partial \mathbf{r}}{\partial r}}{\left| \frac{\partial \mathbf{r}}{\partial r} \right|} \qquad (5.229)$$

Substituting Eqs. (5.223) and (5.224) into Eq. (5.229) yields,

$$\hat{r} = \sin \theta \cos \phi \mathbf{a}_x + \sin \theta \sin \phi \mathbf{a}_y + \cos \theta \mathbf{a}_z \qquad (5.230)$$

The unit vector $\mathbf{a}_\theta$ is defined as,

$$\hat{\theta} = \mathbf{a}_\theta = \frac{\frac{\partial \mathbf{r}}{\partial \theta}}{\left| \frac{\partial \mathbf{r}}{\partial \theta} \right|} \qquad (5.231)$$

Substituting Eqs. (5.225) and (5.226) into Eq. (5.231) yields,

$$\hat{\theta} = \cos \theta \cos \phi \mathbf{a}_x + \cos \theta \sin \phi \mathbf{a}_y - \sin \theta \mathbf{a}_z \qquad (5.232)$$

The unit vector $\mathbf{a}_\phi$ is defined as,

$$\hat{\phi} = \mathbf{a}_\phi = \frac{\frac{\partial \mathbf{r}}{\partial \phi}}{\left| \frac{\partial \mathbf{r}}{\partial \phi} \right|} \qquad (5.233)$$

Substituting Eqs. (5.227) and (5.228) into Eq. (5.233) yields,

$$\hat{\phi} = -\sin \phi \mathbf{a}_x + \cos \phi \mathbf{a}_y \qquad (5.234)$$

Differentiating Eq. (5.230) with respect to r, θ and ϕ yields,

$$\frac{\partial \hat{r}}{\partial r} = 0 \tag{5.235}$$

$$\frac{\partial \hat{r}}{\partial \theta} = \cos\theta \cos\phi \, \mathbf{a}_x + \cos\theta \sin\phi \, \mathbf{a}_y - \sin\theta \mathbf{a}_z = \hat{\theta} \tag{5.236}$$

$$\frac{\partial \hat{r}}{\partial \phi} = -\sin\theta \sin\phi \, \mathbf{a}_x + \sin\theta \cos\phi \, \mathbf{a}_y = \sin\theta.(\hat{\phi}) \tag{5.237}$$

Differentiating Eq. (5.232) with respect to r, θ and ϕ yields,

$$\frac{\partial \hat{\theta}}{\partial r} = 0 \tag{5.238}$$

$$\frac{\partial \hat{\theta}}{\partial \theta} = -\sin\theta \cos\phi \, \mathbf{a}_x - \sin\theta \sin\phi \, \mathbf{a}_y - \cos\theta \mathbf{a}_z = -\hat{r} \tag{5.239}$$

$$\frac{\partial \hat{\theta}}{\partial \phi} = -\cos\theta \sin\phi \, \mathbf{a}_x + \cos\theta \cos\phi \, \mathbf{a}_y = \cos\theta.(\hat{\phi}) \tag{5.240}$$

Differentiating Eq. (5.234) with respect to r, θ and ϕ yields,

$$\frac{\partial \hat{\phi}}{\partial r} = 0 \tag{5.241}$$

$$\frac{\partial \hat{\phi}}{\partial \theta} = 0 \tag{5.242}$$

$$\frac{\partial \hat{\phi}}{\partial \phi} = -\cos\phi \, \mathbf{a}_x - \sin\phi \, \mathbf{a}_y \tag{5.243}$$

The Del operator in spherical coordinates can be written as,

$$\nabla = \frac{\hat{r}}{n_r} \frac{\partial}{\partial r} + \frac{\hat{\theta}}{n_\theta} \frac{\partial}{\partial \theta} + \frac{\hat{\phi}}{n_\phi} \frac{\partial}{\partial \phi} \tag{5.244}$$

$$\nabla^2 = \nabla.\nabla = \left(\frac{\hat{r}}{n_r} \frac{\partial}{\partial r} + \frac{\hat{\theta}}{n_\theta} \frac{\partial}{\partial \theta} + \frac{\hat{\phi}}{n_\phi} \frac{\partial}{\partial \phi} \right).\nabla \tag{5.245}$$

$$\nabla^2 = \frac{\hat{r}}{n_r} \frac{\partial}{\partial r}.\nabla + \frac{\hat{\theta}}{n_\theta} \frac{\partial}{\partial \theta}.\nabla + \frac{\hat{\phi}}{n_\phi} \frac{\partial}{\partial \phi}.\nabla \tag{5.246}$$

$$\nabla^2 = A_1 + A_2 + A_3 \tag{5.247}$$

$$A_1 = \frac{\hat{r}}{n_r}\frac{\partial}{\partial r}\cdot \nabla = \frac{\hat{r}}{n_r}\frac{\partial}{\partial r}\cdot\left(\frac{\hat{r}}{n_r}\frac{\partial}{\partial r} + \frac{\hat{\theta}}{n_\theta}\frac{\partial}{\partial \theta} + \frac{\hat{\phi}}{n_\phi}\frac{\partial}{\partial \phi}\right) \tag{5.248}$$

$$A_1 = \frac{\hat{r}}{n_r}\frac{\partial}{\partial r}\cdot\left(\frac{\hat{r}}{1}\frac{\partial}{\partial r} + \frac{\hat{\theta}}{r}\frac{\partial}{\partial \theta} + \frac{\hat{\phi}}{r\sin\theta}\frac{\partial}{\partial \phi}\right) \tag{5.249}$$

$$A_1 = (\hat{r}.\hat{r})\frac{\partial^2}{\partial r^2} + \hat{r}\frac{\partial \hat{r}}{\partial r}\frac{\partial}{\partial r} + \frac{1}{r}(\hat{r}.\hat{\theta})\frac{\partial^2}{\partial r\partial\theta} + \frac{\hat{r}}{r}(\hat{r}.\hat{\theta})\frac{\partial\hat{\theta}}{\partial r}\frac{\partial}{\partial\theta} + \frac{\hat{r}}{1}(\hat{r}.\hat{\theta})\frac{\partial(1/r)}{\partial r}\frac{\partial}{\partial\theta}$$
$$+ \frac{(\hat{r}.\hat{\phi})}{\sin\theta}\frac{\partial(1/r)}{\partial r}\frac{\partial}{\partial\phi} + \frac{(\hat{r}.\hat{\phi})}{r\sin\theta}\frac{\partial^2}{\partial r\partial\phi} + \frac{\hat{r}}{r\sin\theta}\frac{\partial\hat{\phi}}{\partial r}\frac{\partial}{\partial\phi} \tag{5.250}$$

Substituting Eqs. (5.235), (5.238) and (5.242) and properties of dot products into (5.250) yields,

$$A_1 = \frac{\partial^2}{\partial r^2} \tag{5.251}$$

$$A_2 = \frac{\hat{\theta}}{n_\theta}\frac{\partial}{\partial\theta}\cdot\nabla \tag{5.252}$$

$$A_2 = \frac{\hat{\theta}}{n_\theta}\frac{\partial}{\partial\theta}\cdot\left(\frac{\hat{r}}{1}\frac{\partial}{\partial r} + \frac{\hat{\theta}}{r}\frac{\partial}{\partial\theta} + \frac{\hat{\phi}}{r\sin\theta}\frac{\partial}{\partial\phi}\right) \tag{5.253}$$

$$A_2 = \left(\frac{\hat{\theta}.\hat{r}}{r}\right)\frac{\partial^2}{\partial r\partial\theta} + \frac{\hat{\theta}}{r}\frac{\partial\hat{r}}{\partial\theta}\frac{\partial}{\partial r} + \frac{1}{r^2}(\hat{\theta}.\hat{\theta})\frac{\partial^2}{\partial\theta^2} + \frac{\hat{\theta}}{r^2}\frac{\partial\hat{\theta}}{\partial\theta}\frac{\partial}{\partial\theta} + + \frac{(\hat{\theta}.\hat{\theta})}{r}\frac{\partial(1/r)}{\partial\theta}\frac{\partial}{\partial\theta}$$
$$+ \frac{(\hat{r}.\hat{\phi})}{r}\frac{1}{r\sin\theta}\frac{\partial^2}{\partial\theta\partial\phi} + \frac{(\hat{\theta}.\hat{\phi})}{r}\frac{\partial(r\sin\theta^{-1})}{\partial\theta}\frac{\partial}{\partial\phi} + \frac{\hat{\theta}}{r^2\sin\theta}\frac{\partial\hat{\phi}}{\partial\theta}\frac{\partial}{\partial\phi} \tag{5.254}$$

Substituting Eqs. (5.236), (5.239) and (5.241) and properties of dot products into (5.254) yields,

$$A_2 = \frac{\hat{\theta}}{r}\hat{\theta}\frac{\partial}{\partial r} + \frac{1}{r^2}\frac{\partial^2}{\partial\theta^2} \tag{5.255}$$

$$A_2 = \frac{1}{r}\frac{\partial}{\partial r} + \frac{1}{r^2}\frac{\partial^2}{\partial\theta^2} \tag{5.256}$$

$$A_3 = \frac{\hat{\phi}}{n_\phi}\frac{\partial}{\partial\phi}.\nabla = \frac{\hat{\phi}}{n_\phi}\frac{\partial}{\partial\phi}.\left(\frac{\hat{r}}{n_r}\frac{\partial}{\partial r} + \frac{\hat{\theta}}{n_\theta}\frac{\partial}{\partial\theta} + \frac{\hat{\phi}}{n_\phi}\frac{\partial}{\partial\phi}\right) \tag{5.257}$$

$$A_3 = \frac{\hat{\phi}}{r\sin\theta}\frac{\partial}{\partial\phi}.\left(\frac{\hat{r}}{1}\frac{\partial}{\partial r} + \frac{\hat{\theta}}{r}\frac{\partial}{\partial\theta} + \frac{\hat{\phi}}{r\sin\theta}\frac{\partial}{\partial\phi}\right) \tag{5.258}$$

$$A_3 = \frac{(\hat{\phi}.\hat{r})}{r\sin\theta}\frac{\partial^2}{\partial r\partial\phi} + \frac{\phi}{r\sin\theta}\frac{\partial\hat{r}}{\partial\phi}\frac{\partial}{\partial r} + \frac{(\hat{\phi}.\hat{\theta})}{r^2\sin\theta}\frac{\partial^2}{\partial\phi\partial\theta} + \frac{(\hat{\phi}.\hat{\theta})}{r\sin\theta}\frac{\partial(1/r)}{\partial\phi}\frac{\partial}{\partial\theta}$$
$$+ \frac{\hat{\phi}}{r^2\sin\theta}\frac{\partial\hat{\theta}}{\partial\phi}\frac{\partial}{\partial\theta} + \frac{(\hat{\phi}.\hat{\phi})}{(r\sin\theta)^2}\frac{\partial^2}{\partial\phi^2} + \frac{(\hat{\phi}.\hat{\phi})}{r\sin\theta}\frac{\partial(r\sin\theta^{-1})}{\partial\phi}\frac{\partial}{\partial\phi} + \frac{\hat{\phi}}{(r\sin\theta)^2}\frac{\partial\hat{\phi}}{\partial\phi}\frac{\partial}{\partial\phi} \tag{5.259}$$

The dot product of Eqs. (5.234) and (5.243) provides,

$$\hat{\phi}.\frac{\partial\hat{\phi}}{\partial\phi} = \sin\phi\cos\phi - \sin\phi\cos\phi = 0 \tag{5.260}$$

Substituting Eqs. (5.237), (5.240) and (5.260) and properties of dot products into Eq. (5.259) yields,

$$A_3 = \frac{\hat{\phi}}{r\sin\theta}(\sin\theta.\hat{\phi})\frac{\partial}{\partial r} + \frac{\hat{\phi}}{r^2\sin\theta}(\cos\theta.\hat{\phi})\frac{\partial}{\partial\theta} + \frac{1}{(r\sin\theta)^2}\frac{\partial^2}{\partial\phi^2} \tag{5.261}$$

$$A_3 = \frac{\sin\theta}{r\sin\theta}(\hat{\phi}.\hat{\phi})\frac{\partial}{\partial r} + \frac{\hat{\phi}.\hat{\phi}}{r^2\tan\theta}\frac{\partial}{\partial\theta} + \frac{1}{(r\sin\theta)^2}\frac{\partial^2}{\partial\phi^2} \tag{5.262}$$

$$A_3 = \frac{1}{r}\frac{\partial}{\partial r} + \frac{1}{r^2\tan\theta}\frac{\partial}{\partial\theta} + \frac{1}{(r\sin\theta)^2}\frac{\partial^2}{\partial\phi^2} \tag{5.263}$$

Substituting Eqs. (5.251), (5.256) and (5.263) into Eq. (5.247) yields,

$$\nabla^2 = \frac{\partial^2}{\partial r^2} + \frac{1}{r}\frac{\partial}{\partial r} + \frac{1}{r^2}\frac{\partial^2}{\partial\theta^2} + \frac{1}{r}\frac{\partial}{\partial r} + \frac{1}{r^2\tan\theta}\frac{\partial}{\partial\theta} + \frac{1}{(r\sin\theta)^2}\frac{\partial^2}{\partial\phi^2} \tag{5.264}$$

$$\nabla^2 = \frac{1}{r}\frac{\partial}{\partial r}\left(r^2\frac{\partial}{\partial r}\right) + \frac{1}{r^2\sin\theta}\frac{\partial}{\partial\theta}\left(\sin\theta\frac{\partial}{\partial\theta}\right) + \frac{1}{(r\sin\theta)^2}\frac{\partial^2}{\partial\phi^2} \tag{5.265}$$

5.14   Solution of Poisson's Equation

Consider that Poisson's equation varies in the x-direction only in Cartesian coordinates. Then, the Poisson's equation for one dimension can be reduced as,

$$\frac{d^2V}{dx^2} = -\frac{\rho}{\varepsilon} \tag{5.266}$$

Integrating Eq. (5.266) yields,

$$\frac{dV}{dx} = -\frac{\rho}{\varepsilon}x + A \tag{5.267}$$

Again, integrating Eq. (5.267) yields,

$$V = -\frac{\rho}{2\varepsilon}x^2 + Ax + B \tag{5.268}$$

Consider the boundary conditions $V = V_1$ at $x = x_1$ and $V = V_2$ at $x = x_2$. Then in terms of boundary conditions Eq. (5.268) can be written as,

$$V_1 = -\frac{\rho}{2\varepsilon}x_1^{\;2} + Ax_1 + B \tag{5.269}$$

$$V_2 = -\frac{\rho}{2\varepsilon}x_2^{\;2} + Ax_2 + B \tag{5.270}$$

Subtracting Eq. (5.270) from Eq. (5.269) yields,

$$V_1 - V_2 = \frac{\rho}{2\varepsilon}x_2^{\;2} - \frac{\rho}{2\varepsilon}x_1^{\;2} + A(x_1 - x_2) \tag{5.271}$$

$$A(x_1 - x_2) = (V_1 - V_2) + \frac{\rho}{2\varepsilon}(x_1 + x_2)(x_1 - x_2) \tag{5.272}$$

$$A = \frac{V_1 - V_2}{x_1 - x_2} + \frac{\rho}{2\varepsilon}(x_1 + x_2) \tag{5.273}$$

Substituting Eq. (5.273) into Eq. (5.268) yields,

$$V_1 = -\frac{\rho}{2\varepsilon}x_1^{\;2} + \left[\frac{V_1 - V_2}{x_1 - x_2} + \frac{\rho}{2\varepsilon}(x_1 + x_2)\right]x_1 + B \tag{5.274}$$

$$B = V_1 - \frac{V_1 - V_2}{x_1 - x_2}x_1 - \frac{\rho}{2\varepsilon}x_1^{\;2} + \frac{\rho}{2\varepsilon}x_1^{\;2} - \frac{\rho}{2\varepsilon}x_1x_2 \tag{5.275}$$

$$B = \frac{V_2 x_1 - V_1 x_2}{x_1 - x_2} - \frac{\rho}{2\varepsilon} x_1 x_2 \qquad (5.276)$$

Substituting Eqs. (5.273) and (5.276) into Eq. (5.268) yields,

$$V = -\frac{\rho}{2\varepsilon} x^2 + \left[\frac{V_1 - V_2}{x_1 - x_2} + \frac{\rho}{2\varepsilon}(x_1 + x_2) \right] x + \frac{V_2 x_1 - V_1 x_2}{x_1 - x_2} - \frac{\rho}{2\varepsilon} x_1 x_2 \qquad (5.277)$$

5.15   Numerical Solution of Laplace's Equation

There are few numerical methods that are generally used to find electric potential and field distribution of a specific object, such as, grounding grid, earth electrode, in the area of electrical engineering. These are finite difference method, finite element method, and boundary element method. The available commercial software in this area is developed based on Laplace's and Poisson's equations. In the finite difference method (FDM), the selected object is divided into forward and backward directions with an equal length; for example, a two-dimensional square mesh object is shown in Fig. 5.17. Consider that the length of each side is h and potentials of the points 0, 1, 2, 3 and 4 are V_0, V_1, V_2, V_3 and V_4, respectively. In this case, the voltage does not vary in the z-direction. Therefore, the Laplace equation in two-dimension is [5, 6],

$$\frac{\partial^2 V}{\partial x^2} + \frac{\partial^2 V}{\partial y^2} = 0 \qquad (5.278)$$

For the $x-$ axis, the voltage derivative in the forward direction is,

$$\left. \frac{\partial V}{\partial x} \right|_a = \frac{V_1 - V_0}{h} \qquad (5.279)$$

Fig. 5.17 A square mesh object

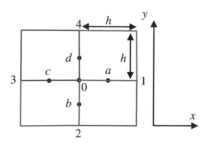

In the backward direction, the voltage derivative is,

$$\left. \frac{\partial V}{\partial x} \right|_c = \frac{V_0 - V_3}{h} \tag{5.280}$$

According to rules of derivative calculus, the following equation can be written:

$$\left. \frac{\partial^2 V}{\partial x^2} \right|_0 = \frac{\left. \frac{\partial V}{\partial x} \right|_a - \left. \frac{\partial V}{\partial x} \right|_c}{h} \tag{5.281}$$

Substituting Eqs. (5.279) and (5.280) into Eq. (5.281) yields,

$$\left. \frac{\partial^2 V}{\partial x^2} \right|_0 = \frac{V_1 - V_0 - V_0 + V_3}{h^2} \tag{5.282}$$

For the y-axis, the voltage derivative in the forward direction is,

$$\left. \frac{\partial V}{\partial y} \right|_d = \frac{V_4 - V_0}{h} \tag{5.283}$$

The voltage derivative in the backward direction is,

$$\left. \frac{\partial V}{\partial y} \right|_b = \frac{V_0 - V_2}{h} \tag{5.284}$$

Again, applying rules of derivative calculus, the following equation can be written:

$$\left. \frac{\partial^2 V}{\partial y^2} \right|_0 = \frac{\left. \frac{\partial V}{\partial y} \right|_d - \left. \frac{\partial V}{\partial y} \right|_b}{h} \tag{5.285}$$

Substituting Eqs. (5.283) and (5.284) into Eq. (5.285) yields,

$$\left. \frac{\partial^2 V}{\partial y^2} \right|_0 = \frac{V_4 - V_0 - V_0 + V_2}{h^2} \tag{5.286}$$

Substituting Eqs. (5.282) and (5.267) into the Eq. (5.278) yields,

$$\frac{V_1 - V_0 - V_0 + V_3}{h^2} + \frac{V_4 - V_0 - V_0 + V_2}{h^2} = 0 \tag{5.287}$$

$$4V_0 = V_1 + V_2 + V_3 + V_4 \tag{5.288}$$

$$V_0 = \frac{V_1 + V_2 + V_3 + V_4}{4} \tag{5.289}$$

The value of the potential V_0 can be found if the potentials at the corners of the mesh are known.

The finite element method is another numerical technique to solve two-dimensional Laplace equations. In 1943, the finite element method (FEM) was first developed by R. Courant to obtain the approximate solution of a complex object. Initially, this technique was applied in mechanical and civil engineering fields to study respective parameters. Later on, the finite element method started being used in the electrical engineering domain to find the flux, potential and electric field distributions around and inside an object. The selected object is discretized either by triangular or rectangular elements. According to P.P. Silvester and R.L. Ferrari the approximate solution of potential for the whole region is,

$$V(x, y) = \sum_{e=1}^{N} V_e(x, y) \tag{5.290}$$

where e represents the number of the elements and N is the total number of triangular elements. The polynomial approximation for V_e within a single element is,

$$V_e(x, y) = a + bx + cy \tag{5.291}$$

The electric field within the element is,

$$\mathbf{E}_e = -\nabla V_e \tag{5.292}$$

Substituting Eq. (5.291) into Eq. (5.292) yields,

$$\mathbf{E}_e = -(b\mathbf{a}_x + c\mathbf{a}_y) \tag{5.293}$$

The potentials V_{e1}, V_{e2} and V_{e3} at nodes 1, 2 and 3 of the triangular element as shown in Fig. 5.18 is,

Fig. 5.18 A triangular element

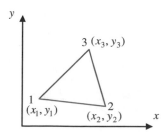

$$V_{e1} = a + bx_1 + cy_1 \tag{5.294}$$

$$V_{e2} = a + bx_2 + cy_2 \tag{5.295}$$

$$V_{e3} = a + bx_3 + cy_3 \tag{5.296}$$

Equations (5.294)–(5.296) can be arranged in the matrix format as,

$$\begin{bmatrix} V_{e1} \\ V_{e2} \\ V_{e3} \end{bmatrix} = \begin{bmatrix} 1 & x_1 & y_1 \\ 1 & x_2 & y_2 \\ 1 & x_3 & y_3 \end{bmatrix} \begin{bmatrix} a \\ b \\ c \end{bmatrix} \tag{5.297}$$

From Eq. (5.297) the coefficients a, b and c can be determined as,

$$\begin{bmatrix} a \\ b \\ c \end{bmatrix} = \begin{bmatrix} 1 & x_1 & y_1 \\ 1 & x_2 & y_2 \\ 1 & x_3 & y_3 \end{bmatrix}^{-1} \begin{bmatrix} V_{e1} \\ V_{e2} \\ V_{e3} \end{bmatrix} \tag{5.298}$$

$$\begin{bmatrix} a \\ b \\ c \end{bmatrix} = \frac{1}{2A} \begin{bmatrix} x_2y_3 - x_3y_2 & x_3y_1 - x_1y_3 & x_1y_2 - x_2y_1 \\ y_2 - y_3 & y_3 - y_1 & y_1 - y_2 \\ x_3 - x_2 & x_1 - x_3 & x_2 - x_1 \end{bmatrix} \begin{bmatrix} V_{e1} \\ V_{e2} \\ V_{e3} \end{bmatrix} \tag{5.299}$$

where A is the area of the element and it can be written as,

$$A = \frac{1}{2} \begin{bmatrix} 1 & x_1 & y_1 \\ 1 & x_2 & y_2 \\ 1 & x_3 & y_3 \end{bmatrix} \tag{5.300}$$

Substituting Eq. (5.299) into Eq. (5.291) yields,

$$V_e(x, y) = \begin{bmatrix} 1 & x & y \end{bmatrix} \frac{1}{2A} \begin{bmatrix} x_2y_3 - x_3y_2 & x_3y_1 - x_1y_3 & x_1y_2 - x_2y_1 \\ y_2 - y_3 & y_3 - y_1 & y_1 - y_2 \\ x_3 - x_2 & x_1 - x_3 & x_2 - x_1 \end{bmatrix} \begin{bmatrix} V_{e1} \\ V_{e2} \\ V_{e3} \end{bmatrix} \tag{5.301}$$

$$V_e(x, y) = \begin{bmatrix} \alpha_1 \\ \alpha_2 \\ \alpha_3 \end{bmatrix} \begin{bmatrix} V_{e1} \\ V_{e2} \\ V_{e3} \end{bmatrix} \tag{5.302}$$

$$V_e(x, y) = \sum_{i=1}^{3} \alpha_i(x, y) V_{ei} \tag{5.303}$$

Here α_1, α_2 and α_3 are the shape functions of the element and their expressions from the Eq. (5.301) can be written as,

$$\alpha_1 = \frac{1}{2A}[(x_2y_3 - x_3y_2) + (y_2 - y_3)x + (x_3 - x_2)y] \qquad (5.304)$$

$$\alpha_2 = \frac{1}{2A}[(x_3y_1 - x_1y_3) + (y_3 - y_1)x + (x_1 - x_3)y] \qquad (5.305)$$

$$\alpha_3 = \frac{1}{2A}[(x_1y_2 - x_2y_1) + (y_1 - y_2)x + (x_2 - x_1)y] \qquad (5.306)$$

The energy per unit length of an element is given by,

$$W_e = \frac{1}{2}\int_s \varepsilon|\mathbf{E}|^2 dS \qquad (5.307)$$

Substituting $\mathbf{E} = -\nabla V_e$ into Eq. (5.307) yields,

$$W_e = \frac{1}{2}\int_s \varepsilon|\nabla V_e|^2 dS \qquad (5.308)$$

Equation (5.303) can be modified as,

$$\nabla V_e = \sum_{i=1}^{3} V_{ei}\nabla\alpha_i \qquad (5.309)$$

Substituting Eq. (5.129) into Eq. (5.308) provides,

$$W_e = \frac{1}{2}\sum_{i=1}^{3}\sum_{j=1}^{3}\varepsilon V_{ei}V_{ej}\left[\int_s \nabla\alpha_i.\nabla\alpha_j dS\right] \qquad (5.310)$$

$$W_e = \frac{1}{2}\sum_{i=1}^{3}\sum_{j=1}^{3}\varepsilon V_{ei}V_{ej}C_{ij} \qquad (5.311)$$

where,

$$C_{ij} = \int_s \nabla\alpha_i.\nabla\alpha_j \, dS \qquad (5.312)$$

Equation (5.312) can be written in the matrix format as,

$$W_e = \frac{1}{2}\varepsilon[V_e]^T[C][V_e] \qquad (5.313)$$

Where the element coefficient matrix and potentials are,

$$[C] = \begin{bmatrix} C_{11} & C_{12} & C_{13} \\ C_{21} & C_{22} & C_{23} \\ C_{31} & C_{32} & C_{33} \end{bmatrix} \tag{5.314}$$

$$[V_e] = \begin{bmatrix} V_{e1} \\ V_{e2} \\ V_{e3} \end{bmatrix} \tag{5.315}$$

The total energy for all elements can be determined as,

$$W = \sum_{e=1}^{N} W_e = \frac{1}{2}\varepsilon[V]^T[C][V] \tag{5.316}$$

The Laplace equation is satisfied when the total energy in the region is minimum. In this case, it can be expressed as,

$$\frac{\partial W}{\partial V_k} = 0 \quad k = 1, 2, 3, \ldots, n \tag{5.317}$$

For free and prescribed potentials, Eq. (5.316) can be written as,

$$W = \frac{1}{2}\varepsilon[\,V_f \quad V_p\,]\begin{bmatrix} C_{ff} & C_{fp} \\ C_{pf} & C_{pp} \end{bmatrix}\begin{bmatrix} V_f \\ V_p \end{bmatrix} \tag{5.318}$$

Applying Eqs. (5.317) and (5.318) i.e. differentiating it with respect to V_f yields,

$$C_{ff}V_f + C_{fp}V_p = 0 \tag{5.319}$$

$$[C_{ff}][V_f] = -[C_{fp}][V_p] \tag{5.320}$$

$$[A][V] = [B] \tag{5.321}$$

where,

$$[A] = [C_{ff}] \tag{5.322}$$

$$[V] = [V_f] \tag{5.323}$$

Equation (5.321) can be rearranged as,

$$[V] = [A]^{-1}[B] \tag{5.324}$$

Therefore, the potential can be determined from Eq. (5.324) if other parameters are known.

Exercise Problems

5.1 The expression of electric potential in Cartesian coordinates is $V(x, y, z) = x^2 y - z^2 + 8$. Determine the (i) numerical value of the voltage at point $P(1, -1, 2)$, (ii) electric field, and (iii) verify the Laplace equation.

5.2 The electric potential in Cartesian coordinates is given by $V(x, y, z) = e^x - e^{-y} + z^2$. Determine the (i) numerical value of the voltage at point $P(1, 1, -2)$, (ii) electric field at the point $P(1, 1, -2)$, and (iii) verify the Laplace equation.

5.3 The expression of electric potential in cylindrical coordinates is given as $V(\rho, \phi, z) = \rho^2 z \cos \phi$. Determine the (i) numerical value of the voltage at point $P(\rho = -1, \phi = 45°, z = 5)$, (ii) electric field at the point $P(\rho = -1, \phi = 45°, z = 5)$, and (iii) verify the Laplace equation.

5.4 The electric potential in spherical coordinates is given by $V(r, \theta, \phi) = 5r^2 \sin \theta \cos \phi$. Determine the (i) numerical value of the voltage at point $P(r = 1, \theta = 40°, \phi = 120°)$, (ii) electric field at the point $P(r = 1, \theta = 40°, \phi = 120°)$, and (iii) verify the Laplace equation.

5.5 In Cartesian coordinates, the volume charge density is $\rho_v = -1.6 \times 10^{-11} \varepsilon_0 x$ C/m^3 in the free space. Consider $V = 0$ at $x = 0$ and $V = 4$ V at $x = 2$ m. Determine the electric potential and field at $x = 5$ m.

5.6 The charge density in cylindrical coordinates is $\rho_v = \frac{25}{\rho}$ pC/m^3. Consider $V = 0$ at $\rho = 2$ m and $V = 120$ V at $\rho = 5$ m. Calculate the electric potential and field at $\rho = 6$ m.

5.7 Two concentric spherical shells with radius of $r = 1$ m and $r = 2$ m contain the potentials of $V = 0$ and $V = 80$ V respectively. Find the potential and electric field.

5.8 Determine the potential of a rectangular object of infinite length. Consider $a = b = 1$ m, $V_0 = 50$ V, $x = \frac{3a}{2}$ and $y = \frac{b}{2}$.

References

1. J.N. Cernica, *Geotechnical Engineering: Soil Mechanics*. (Wiley, USA, 1995)
2. B.M. Das, *Principles of Geotechnical Engineering*. (PWS-KENT Publishing Company, USA, 1985)
3. D.K. Cheng, *Fundamentals of Engineering Electromagnetics*, 1st edn. (Prentice-Hall Inc., Upper Saddle River, New Jersey, USA, 1993)

4. M.A. Salam, *Electromagnetic Field Theories for Engineering*, 1st edn. (Springer Publishers, 2014), pp. 1–311
5. P.P. Silvester, R.L. Ferrari, *Finite Elements for Electrical Engineers*, 3rd edn. (Cambridge University Press, UK, 1996)
6. C.R. Paul, S.A. Nasar, *Introduction to Electromagnetic Fields*, 1st edn. (McGraw-Hill Inc., USA, 1982)

Chapter 6
Soil Resistivity Measurement

6.1 Introduction

Different types of soil and its characteristics have been discussed in Chap. 4. The value of the soil resistivity is mainly dependent on its properties. Soil resistivity measurement is an important factor in finding the best location for any grounding system. Based on the measurement results, new grounding systems are installed for the power generating station, substation, transmission tower, distribution pole, telephone exchange, industry, commercial and residential buildings.

Nowadays, most of the power utility companies are installing their transmission lines, water and gas pipelines at the same corridor to reduce the land use. The exact value of soil resistivity is very important for the underground pipelines. These underground pipelines are not always parallel with the transmission lines. Sometimes, it runs with different angles with the transmission lines. Therefore, more soil resistivity measurements are need to be carried out at different places. Generally, lower soil resistivity increases the corrosion of underground pipelines. The corrosion ratings of different soil resistivity [1, 2] are shown in Table 6.1.

Basically, the current flow in the underground pipelines increases the corrosion. The sandy soil has a high soil resistivity so its corrosion rating is low. Whereas clay, silt and garden soils have low resistivity which have a high corrosion rating. In this chapter, different methods of soil resistivity measurement will be discussed.

6.2 Two-Pole Method

In this method, a voltage source is connected between a hemisphere earth electrode and an auxiliary probe or potential probe as shown in Fig. 6.1. An ammeter is connected in series with the voltage source to measure the current. For a specific supply voltage and measured current, the resistance can be determined as,

© Springer Science+Business Media Singapore 2016
Md.A. Salam and Q.M. Rahman, *Power Systems Grounding*,
Power Systems, DOI 10.1007/978-981-10-0446-9_6

Table 6.1 Soil resistivity with corrosion rating

Soil resistivity (Ω-m)	Corrosion rating
>200	Essentially non-corrosive
100–200	Mildly corrosive
50–100	Moderately corrosive
30–50	Corrosive
10–30	Highly corrosive
<10	Extremely corrosive

Fig. 6.1 Earth electrode with potential probe

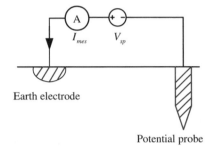

$$R_E = \frac{V_{sp}}{I_{mes}} \tag{6.1}$$

For a hemisphere earth electrode with a radius r, the expression of ground resistance is,

$$R_E = \frac{\rho}{2\pi r} \tag{6.2}$$

Then, the soil resistivity can be determined as,

$$\rho = 2\pi r R_E \tag{6.3}$$

This method is easy to measure the soil resistivity at any small space. In this case, fluke 1625 m can be used to perform the soil resistivity measurement. The connection diagram of two-pole method for the soil resistance measurement is shown in Fig. 6.2.

6.3   Four-Pole Equal Method

In 1915, an American Geologist Dr. Frank Wenner of US Bureau of Standard introduced four-pole equal method to measure the soil resistivity. According to his name, this method is also known as Wenner method. The overall setup for the

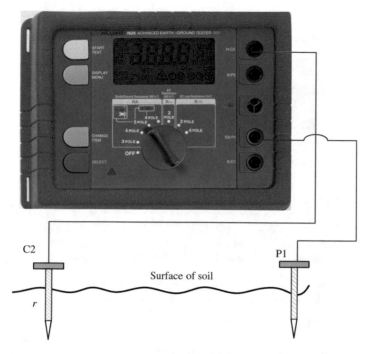

Fig. 6.2 Connection diagram of two-pole method [3]

four-pole equal method is shown in Fig. 6.3. In this method, four equidistant probes are inserted into the soil on a straight line as shown in Fig. 6.3. Then the current terminals (C1 and C2) of the advanced earth testing meter, fluke 1625 are connected to the two outer probes, and the potential probing terminals (P1 and P2) are connected to the two inner probes. Then press the start button of the meter which injects the current into the soil through the current probes and the resulting voltage is measured across the potential probes (inner probes).

According to Ohms law, the meter calculates the soil resistance, and then displays the soil resistance-value. From the measured resistance-value, soil resistivity is calculated using the following formulae,

$$\rho = \frac{4\pi a R_E}{1 + \frac{2a}{\sqrt{a^2 + 4b^2}} - \frac{2a}{\sqrt{2a^2 + 4b^2}}} \tag{6.4}$$

As shown in Fig. 6.3, a is the distance between two adjacent probes, and b is the length of the probe (probe-depth) inserted into the soil. If b (In general, $b = 4a$) is very large compared to a, Eq. (6.4) reduces to,

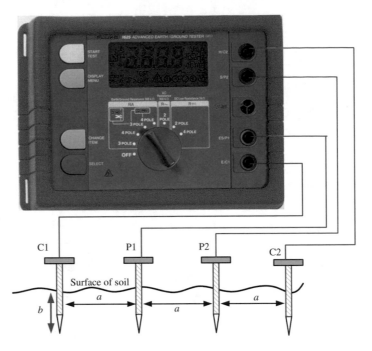

Fig. 6.3 Connection diagram of four poles method [3]

$$\rho = 4\pi a R_E \tag{6.5}$$

If b is very small ($b \ll a$) compared to a, Eq. (6.4) reduces to,

$$\rho = \frac{4\pi a R_E}{1 + 2 - 1} \tag{6.6}$$

$$\rho = 2\pi a R_E \tag{6.7}$$

During the measurement, the distance between the adjacent probes may be considered to have a value between 1 and 50 ft. This worth nothing that this adjacent distance depends on the available free space near the grounding system.

6.4   Derivation of Resistivity

One-half of a sphere is inserted into the soil as shown in Fig. 6.4 where the buried part forms a hemisphere. A current I is inserted into the soil whose resistivity is ρ and it is distributed to the ground. The current density at the surface of the hemisphere is,

Fig. 6.4 Current in the soil through hemisphere

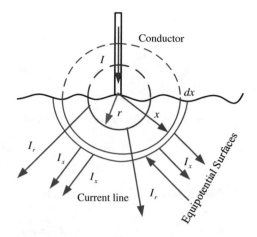

$$J = \frac{1}{A} = \frac{I}{2\pi r^2} \tag{6.8}$$

The current density at any distance x from the center of the hemisphere is,

$$J(x) = \frac{I}{2\pi x^2} \tag{6.9}$$

The electric field at any distance x from the center of the hemisphere can be determined as,

$$E(x) = \rho J(x) \tag{6.10}$$

Substituting Eq. (6.9) into Eq. (6.10) yields,

$$E(x) = \frac{\rho I}{2\pi x^2} \tag{6.11}$$

The potential difference from the center to earth can be determined as,

$$V_x = \int_r^x E(x)dx \tag{6.12}$$

Substituting Eq. (6.11) into Eq. (6.12) yields,

$$V_x = \int_r^x \frac{\rho I}{2\pi x^2} dx \tag{6.13}$$

$$V_x = -\frac{\rho I}{2\pi}\left[\frac{1}{x}\right]_r^x \tag{6.14}$$

$$V_x = -\frac{\rho I}{2\pi}\left[\frac{1}{r} - \frac{1}{x}\right] \tag{6.15}$$

If $x = \infty$, then the expression of potential difference can be modified as,

$$V_x = \frac{\rho I}{2\pi r} \tag{6.16}$$

The expression of the soil resistance is,

$$R_E = \frac{V_{12}}{I} = \frac{\rho}{2\pi r} \tag{6.17}$$

Then the expression of soil resistivity can be written as,

$$\rho = 2\pi r R_E \tag{6.18}$$

Four probes are considered as spheres and they are placed on a straight line with an equal separation distance a as shown in Fig. 6.5. The current enters through the first sphere and is distributed radially into the soil. This current will come out of the sphere 4. The distances of the spheres 2 and 3 from the sphere 1 are a and $2a$, respectively. According to Eq. (6.15), the potential at the sphere 2 due to the current flowing through the sphere 1 can be written as,

$$V_2 = \frac{\rho I}{4\pi}\left[\frac{1}{a} - \frac{1}{2a}\right] \tag{6.19}$$

Similarly, the potential at the sphere 3 due to the current flowing out of the sphere 4 is,

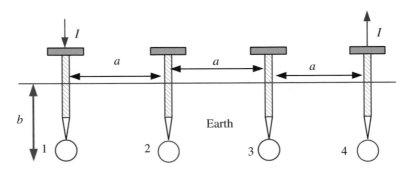

Fig. 6.5 Four spheres in the soil

$$V_3 = \frac{\rho I}{4\pi}\left[\frac{1}{2a} - \frac{1}{a}\right] \tag{6.20}$$

The potential difference between the spheres 2 and 3 is,

$$V = V_2 - V_3 \tag{6.21}$$

Substituting Eqs. (6.19) and (6.20) into Eq. (6.21) yields,

$$V = \frac{\rho I}{4\pi}\left[\frac{1}{a} - \frac{1}{2a}\right] - \frac{\rho I}{4\pi}\left[\frac{1}{2a} - \frac{1}{a}\right] \tag{6.22}$$

$$V = \frac{\rho I}{4\pi}\left[\frac{1}{a} - \frac{1}{2a} - \frac{1}{2a} + \frac{1}{a}\right] \tag{6.23}$$

$$V = \frac{\rho I}{4\pi}\left[\frac{2}{a} - \frac{1}{a}\right] \tag{6.24}$$

$$\rho = 4\pi a \frac{V}{I} = 4\pi a R_E \tag{6.25}$$

Two conductors are placed on the top soil surface and four conductors are placed on a straight line inside the soil that forms the bottom soil surface as shown in Fig. 6.6.

From Fig. 6.6, the following equations can be written as,

$$r_{12} = r_{34} = a \tag{6.26}$$

$$r_{13} = r_{42} = 2a \tag{6.27}$$

$$r_{53} = r_{62} = \sqrt{4a^2 + 4b^2} \tag{6.28}$$

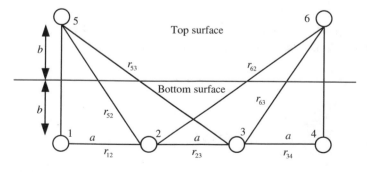

Fig. 6.6 Conductors at the top and bottom surface of the soil

$$r_{52} = r_{63} = \sqrt{a^2 + 4b^2} \tag{6.29}$$

Initially, consider the current enters into the conductor 1 and leaving out of the conductor 4. The potentials at the conductors 2 and 3 are V_a and V_b, respectively. According to Eq. (6.15), the expression of these potentials can be written as,

$$V_a = \frac{\rho I}{4\pi} \left[\frac{1}{r_{12}} - \frac{1}{r_{24}} \right] \tag{6.30}$$

$$V_b = \frac{\rho I}{4\pi} \left[\frac{1}{r_{13}} - \frac{1}{r_{43}} \right] \tag{6.31}$$

Substituting Eqs. (6.26) and (6.27) into Eqs. (6.30) and (6.31) yields,

$$V_a = \frac{\rho I}{4\pi} \left[\frac{1}{a} - \frac{1}{2a} \right] = \frac{\rho I}{4\pi} \frac{1}{2a} \tag{6.32}$$

$$V_b = \frac{\rho I}{4\pi} \left[\frac{1}{2a} - \frac{1}{a} \right] = \frac{\rho I}{4\pi} \left(-\frac{1}{2a} \right) \tag{6.33}$$

The potential differences between the conductors 2 and 3 is,

$$V_m = V_a - V_b \tag{6.34}$$

Substituting Eqs. (6.32) and (6.33) into Eq. (6.34) yields,

$$V_m = \frac{\rho I}{4\pi} \frac{1}{2a} - \frac{\rho I}{4\pi} \left(-\frac{1}{2a} \right) \tag{6.35}$$

$$V_m = \frac{\rho I}{4\pi} \left(\frac{1}{2a} + \frac{1}{2a} \right) \tag{6.36}$$

$$V_m = \frac{\rho I}{4\pi} \frac{1}{a} \tag{6.37}$$

Again, consider the current is entering in the conductor 5 and leaving out of the conductor 6. In this case, the potentials at the conductors 2 and 3 are,

$$V_c = \frac{\rho I}{4\pi} \left[\frac{1}{r_{52}} - \frac{1}{r_{62}} \right] \tag{6.38}$$

$$V_d = \frac{\rho I}{4\pi} \left[\frac{1}{r_{53}} - \frac{1}{r_{63}} \right] \tag{6.39}$$

Again, the potential difference between the conductors 2 and 3 is,

$$V_n = V_c - V_d \tag{6.40}$$

Substituting Eqs. (6.38) and (6.39) into Eq. (6.40) yields,

$$V_n = \frac{\rho I}{4\pi} \left[\frac{1}{r_{52}} - \frac{1}{r_{62}} - \frac{1}{r_{53}} + \frac{1}{r_{63}} \right] \tag{6.41}$$

Substituting Eqs. (6.28) and (6.29) into Eqs. (6.41) yields,

$$V_n = \frac{\rho I}{4\pi} \left[\frac{1}{\sqrt{a^2 + 4b^2}} - \frac{1}{\sqrt{4a^2 + 4b^2}} - \frac{1}{\sqrt{4a^2 + 4b^2}} + \frac{1}{\sqrt{a^2 + 4b^2}} \right] \tag{6.42}$$

$$V_n = \frac{\rho I}{4\pi} \left[\frac{2}{\sqrt{a^2 + 4b^2}} - \frac{2}{\sqrt{4a^2 + 4b^2}} \right] \tag{6.43}$$

Total potential difference for both cases is,

$$V = V_m + V_n \tag{6.44}$$

Substituting Eqs. (6.37) and (6.43) into Eq. (6.44) yields,

$$V = \frac{\rho I}{4\pi} \left[\frac{1}{a} + \frac{2}{\sqrt{a^2 + 4b^2}} - \frac{2}{\sqrt{4a^2 + 4b^2}} \right] \tag{6.45}$$

$$V = \frac{\rho I}{4\pi a} \left[1 + \frac{2a}{\sqrt{a^2 + 4b^2}} - \frac{2a}{\sqrt{4a^2 + 4b^2}} \right] \tag{6.46}$$

$$\rho = \frac{V}{I} \frac{4\pi a}{\left(1 + \dfrac{2a}{\sqrt{a^2 + 4b^2}} - \dfrac{2a}{\sqrt{4a^2 + 4b^2}} \right)} \tag{6.47}$$

$$\rho = \frac{4\pi a R_E}{1 + \dfrac{2a}{\sqrt{a^2 + 4b^2}} - \dfrac{2a}{\sqrt{4a^2 + 4b^2}}} \tag{6.48}$$

Equation (6.48) can be modified with respect to different relationships between a and b as discussed earlier.

Example 6.1 A soil resistivity measurement is carried out near a power station using the Wenner four poles equal method. The six readings of soil resistance were taken during the measurement using the Fluke 1625 earth tester equipment. The readings are recorded at 1, 2, 3, 4, 5 and 6 m intervals of the probe distance. The corresponding soil resistance were measured to be 35, 18, 13, 11.2, 10.2 and 13 Ω, respectively. Determine the soil resistivity and plot it with respect to the probe distance.

Solution

The value of the soil resistivity can be calculated as,

$$\rho = 2\pi a R_e$$
$$\rho_1 = 2\pi \times 1 \times 35 = 219.8\,\Omega\text{-m}$$
$$\rho_2 = 2\pi \times 2 \times 18 = 226.08\,\Omega\text{-m}$$
$$\rho_3 = 2\pi \times 3 \times 13 = 244.92\,\Omega\text{-m}$$
$$\rho_4 = 2\pi \times 4 \times 11.2 = 281.34\,\Omega\text{-m}$$
$$\rho_5 = 2\pi \times 5 \times 10.2 = 320.28\,\Omega\text{-m}$$
$$\rho_6 = 2\pi \times 6 \times 13 = 489.84\,\Omega\text{-m}$$

The plot of soil resistivity with the probe distance is shown in Fig. 6.7.

Example 6.2 F. Wenner four poles equal method is used to measure the soil resistivity near a 66/11 kV substation using a Fluke 1625 earth tester. The readings are recorded at 1, 2, 3, 4 and 5 m intervals of the probe distance. The corresponding soil resistance were measured to be 16.4, 5.29, 3.05, 1.96 and 1.36 Ω, respectively. Calculate the average soil resistivity in that substation.

Fig. 6.7 Variation of soil resistivity with probe istance

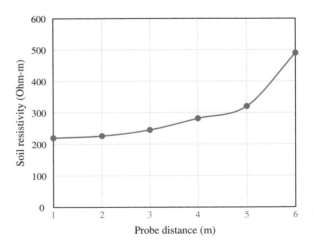

Solution

The value of the soil resistivity can be calculated as,

$$\rho = 2\pi a R_e$$
$$\rho_1 = 2\pi \times 1 \times 16.4 = 103\,\Omega\text{-m}$$
$$\rho_2 = 2\pi \times 2 \times 5.29 = 66.5\,\Omega\text{-m}$$
$$\rho_3 = 2\pi \times 3 \times 3.05 = 57.5\,\Omega\text{-m}$$
$$\rho_4 = 2\pi \times 4 \times 1.96 = 49.3\,\Omega\text{-m}$$
$$\rho_5 = 2\pi \times 5 \times 1.26 = 42.7\,\Omega\text{-m}$$

The average value of the soil resistivity can be determined as,

$$\rho_{av} = \frac{\rho_1 + \rho_2 + \rho_3 + \rho_4 + \rho_5}{5} = \frac{103 + 66.5 + 57.5 + 49.3 + 42.7}{5} = 63.8\,\Omega\text{-m}$$

Practice problem 6.1
The soil resistivity measurement is carried out near a power station at 1, 2, 3, 4, 5 and 6 m intervals of the probe distance using F. Wenner four poles equal method. The corresponding soil resistance were measured to be 16, 3.5, 2.6, 2.01, 1.56 and 1.02 Ω, respectively. Find the soil resistivity and plot it with respect to the probe distance.

Practice problem 6.2
F. Wenner four poles equal method is used to measure the soil resistivity near an 11/69 kV power station using a Fluke 1625 earth tester. The readings are recorded at 1, 2, 3, 4 and 5 m intervals of the probe distance. The corresponding soil resistance were measured to be 12.4, 3.25, 2.95, 1.86 and 0.98 Ω, respectively. Determine the average soil resistivity in that substation.

6.5   Lee's Partitioning Method

Lee introduced a method to measure the soil resistivity by partitioning the potential probes. According to his name, this method is known as Lee's partitioning method. In this method, five probes are used on a straight line as shown in Fig. 6.8. The current enters into the first probe and coming out through the fifth probe. In each measurement, four probes are used. The potentials V_2 and V_3, at the probes 2 and 3 are,

$$V_2 = \frac{\rho I}{2\pi}\left[\frac{1}{a} - \frac{1}{2a}\right] = \frac{\rho I}{4\pi a} \qquad (6.49)$$

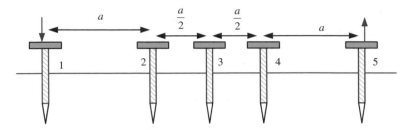

Fig. 6.8 Schematic of Lee's method

$$V_3 = \frac{\rho I}{2\pi}\left[\frac{1}{a+0.5a} - \frac{1}{a+0.5a}\right] = 0 \tag{6.50}$$

The potential difference between the probes 2 and 3 is,

$$V_{23} = V_2 - V_3 = \frac{\rho I}{4\pi a} - 0 \tag{6.51}$$

$$\rho = 4\pi a \frac{V_{23}}{I} = 4\pi a R_{23} \tag{6.52}$$

Similarly, the potential at probes 4 can be written as,

$$V_4 = \frac{\rho I}{2\pi}\left[\frac{1}{2a} - \frac{1}{a}\right] = \frac{\rho}{2\pi}\left(-\frac{1}{2a}\right) \tag{6.53}$$

The potential difference between the probes 3 and 4 is,

$$V_{34} = V_3 - V_4 = 0 - \frac{\rho}{4\pi}\left(-\frac{1}{2a}\right) \tag{6.54}$$

$$\rho = 4\pi a \frac{V_{34}}{I} = \pi a R_{34} \tag{6.55}$$

If the values of the soil resistivity of the two measurements are the same then the soil is considered to be homogeneous.

Example 6.3 Lee's portioning method is used to measure the soil resistivity of a substation. The resistances are measured to be 3 Ω between the probes 2 and 3, 2.95 Ω between the probes 3 and 4. Determine the soil resistivity if the probe separation distance is 2 m in both cases.

Solution

The value of the soil resistivity can be calculated as,

$$\rho = 4\pi a R_{23} = 4\pi \times 2 \times 3 = 75.4\,\Omega\text{-m}$$
$$\rho = 4\pi a R_{34} = 4\pi \times 2 \times 2.95 = 74.14\,\Omega\text{-m}$$

In this case, the soil is homogeneous.

Practice problem 6.3
The resistances are measured to be 1.54 Ω between the probes 2 and 3, 2.25 Ω between the probes 3 and 4 using Lee's portioning method. Find the soil resistivity if the probe separation distance is 1 m in both cases.

6.6   Sided Probe System

In Wenner equal probe system, all the probes along with the connection wires need to be moved for each measurement. Therefore, this method is laborious. In sided probe system, only two probes need to be moved instead of all four probes. The outer second current probes should be placed far away so that the potential difference between the inner probes can be neglected due to the current coming out through it. The current is entering into probe 1 and coming out of probe 4 as shown in Fig. 6.9.

The potential difference between the probes 2 and 3 due to current entering into probe 1 can be written as,

$$V_{23} = \frac{\rho I}{2\pi}\left[\frac{1}{a} - \frac{1}{a+b}\right] \tag{6.56}$$

$$V_{23} = \frac{\rho I}{2\pi}\left[\frac{a+b-a}{a(a+b)}\right] \tag{6.57}$$

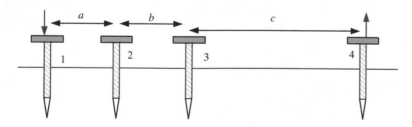

Fig. 6.9 Schematic of sided probe system

$$\rho = 2\pi a \frac{(a+b)}{b} \frac{V_{23}}{I} \qquad (6.58)$$

$$\rho = 2\pi a \frac{(a+b)}{b} R_E \qquad (6.59)$$

The potential difference between the probes 2 and 3 due to current coming out of probe 4 can be written as,

$$V_{23} = \frac{\rho I}{2\pi} \left[\frac{1}{c} - \frac{1}{b+c} \right] \qquad (6.60)$$

$$V_{23} = \frac{\rho I}{2\pi} \left[\frac{b+c-c}{c(b+c)} \right] \qquad (6.61)$$

$$\rho = 2\pi c \frac{(b+c)}{b} \frac{V_{23}}{I} \qquad (6.62)$$

$$\rho = 2\pi c \frac{(b+c)}{b} R_E \qquad (6.63)$$

Dividing Eq. (6.63) by (6.59) yields,

$$1 = \frac{c(b+c)}{a(a+b)} \qquad (6.64)$$

If distances a and c are very large compared to b, then Eq. (6.64) reduces as,

$$1 = \frac{c(c)}{a(a)} \qquad (6.65)$$

$$\frac{c^2}{a^2} = 1 \qquad (6.66)$$

From Eq. (6.66), it is assumed that the ratio of the distance between the potential probe 3 and the current probe 4 to the distance between the current probe 1 and the potential probe 2 is equal or less than 1%. Based on this argument, Eq. (6.66) can be written as,

$$\frac{c^2}{a^2} > 100 \qquad (6.67)$$

$$c > 10a \qquad (6.68)$$

The main drawback of this method is more spatial distance required compared to the Wenner method to complete the soil resistivity measurement.

6.7   Schlumberger Method

In USA and Europe, Schlumberger method was most popular from 1960 to 1990. In this method, two current probes C1 and C2 are placed on the outside and two potential probes P1 and P2 are placed on the inside of the overall setup as shown in Fig. 6.10. The outer current probes need to be moved symmetrically. However, the potential probes are never moved.

Practically, the ratio of the separation distance of potential probes to the separation distance between the current probes should be one fifth or less. In other words, the distance between the current probes should be four or five times the separation distance between the potential probes. Mathematically, it can be written as,

$$\frac{c}{d} \leq \frac{1}{5} \tag{6.69}$$

This method saves some measurement time due to less movement of the probes. However, Wenner method is more straight forward and simpler than the others method. As shown in Fig. 6.11, the fluke 1625 earth tester is used to insert the current into the soil through the current probe C1.

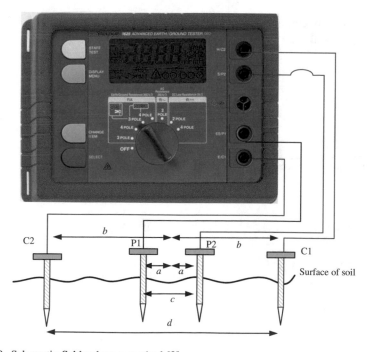

Fig. 6.10 Schematic Schlumberger method [3]

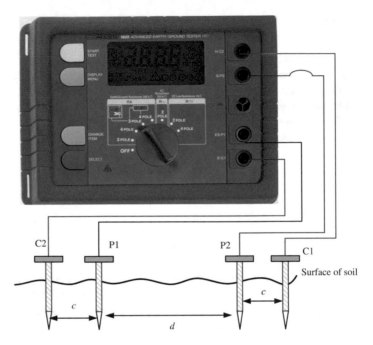

Fig. 6.11 Schematic Schlumberger method for different approach [3]

This current is coming out from the soil through the current probe C2. In this case, the potentials at P1 and P2 are,

$$V_{P1} = \frac{\rho I}{2\pi} \left[\frac{1}{b-a} - \frac{1}{b+a} \right] \tag{6.70}$$

$$V_{P2} = \frac{\rho I}{2\pi} \left[\frac{1}{b+a} - \frac{1}{b-a} \right] \tag{6.71}$$

The potential difference between points P1 and P2 is,

$$V = V_{P1} - V_{P2} \tag{6.72}$$

Substituting Eqs. (6.70) and (6.71) into Eq. (6.72) yields,

$$V = \frac{\rho I}{2\pi} \left[\frac{1}{b-a} - \frac{1}{b+a} - \frac{1}{b+a} + \frac{1}{b-a} \right] \tag{6.73}$$

$$\frac{V}{I} = \frac{4a\rho}{2\pi} \frac{1}{b^2 - a^2} \tag{6.74}$$

$$R_E = \frac{4a\rho}{2\pi} \frac{1}{b^2 - a^2} \tag{6.75}$$

$$\rho = 2\pi R_E \frac{b^2 - a^2}{4a} \tag{6.76}$$

$$\rho = 2\pi a R_E \frac{b^2 - a^2}{4a^2} \tag{6.77}$$

$$\rho = 2\pi a R_E \frac{\frac{b^2}{a^2} - \frac{a^2}{a^2}}{4} \tag{6.78}$$

$$\rho = 2\pi a R_E \frac{\left(\frac{b}{a}\right)^2 - 1}{4} \tag{6.79}$$

$$\rho = 2\pi a R_E \frac{(\alpha)^2 - 1}{4} \tag{6.80}$$

where $\alpha = \frac{b}{a}$. If the value of α is equal to 3, then Eq. (6.80) reduces to the following expression:

$$\rho = 4\pi a R_E \tag{6.81}$$

Equation (6.81) is same as the Wenner equation when the depth of the probe into the soil is very large compared to the adjacent distance between any two probes. The Schlumberger unequal method for soil resistivity measurement can be carried out if there is no physical barrier or obstruction of buried the probes. Figure 6.11 shows the Schlumberger unequal method for alternative expression of soil resistivity. The current is coming out of the soil through the current probe C2. In this case, the expressions of potential at P1 and P2 are,

$$V_{P1} = \frac{\rho I}{2\pi} \left[\frac{1}{c} - \frac{1}{c+d} \right] \tag{6.82}$$

$$V_{P2} = \frac{\rho I}{2\pi} \left[\frac{1}{c+d} - \frac{1}{c} \right] \tag{6.83}$$

Substituting Eqs. (6.82) and (6.83) into Eq. (6.72) yields the potential difference,

$$V = \frac{\rho I}{2\pi} \left[\frac{1}{c} - \frac{1}{c+d} - \frac{1}{c+d} + \frac{1}{c} \right] \tag{6.84}$$

$$V = \frac{\rho I}{2\pi} \left[\frac{2}{c} - \frac{2}{c+d} \right] \tag{6.85}$$

$$V = \frac{2\rho I}{2\pi} \left[\frac{c+d-c}{c(c+d)} \right] \tag{6.86}$$

$$\frac{V}{I} = \frac{\rho}{2\pi} \frac{d}{c(c+d)} \tag{6.87}$$

$$R_E = \frac{\rho}{2\pi} \frac{d}{c(c+d)} \tag{6.88}$$

$$\rho = 2\pi R_E \frac{c(c+d)}{d} \tag{6.89}$$

Example 6.4 The Schlumberger unequal method is used to measure the soil resistivity of a substation. The resistance of the soil is measured to be 6 Ω by setting a distance of 1 m between the current and the potential probes. Calculate the soil resistivity if the two potential probes' separation distance is 2 m.

Solution

The value of the soil resistivity can be determined as,

$$\rho = 2\pi R_E \frac{c(c+d)}{d} = 2\pi \times 6 \times 1 \frac{(1+2)}{2} = 56.55 \, \Omega\text{-m}$$

Practice problem 6.4

The resistance of the soil is measured to be 8 Ω using the Schlumberger unequal method by setting a distance of 1.5 m between the current probe and the potential probe. Determine the soil resistivity if the two potential probes' separation distance is 2 m.

6.8   Different Terms in Grounding System

Different grounding system parameters including ground, grounding, ground current, ground electrode, grounding system, ground resistance, ground impedance, ground grid, ground potential rise, step and touch potentials are discussed below.

Ground: A conducting connection by which an electrical device or component is connected to the earth is known as ground. It is obtained by ground electrode discussed later.

Grounding: It is often known as grounded. A grounding or grounded is a continuous conductive path which can carry any magnitudes of fault currents.

Ground current: The value of the current that either flow into the soil or out of the soil is known as ground current.

Ground electrode: It is copper rod or plate which is driven into the earth or soil to provide a reliable conductive path to the ground. It is also known as earth electrode.

Grounding system: The combination of the ground electrode, ground plate, clamps, ground clips and connecting conductors is known as grounding system.

Ground resistance: The impedance (Z_R) between a ground electrode, ground plate and the remote earth is known as the ground resistance (R_g).

$$Z_R = R_g + j0 \qquad (6.90)$$

Ground impedance: The phasor sum of the resistance (R_g) and the reactance (X_g) between a ground electrode, ground plate and the remote earth is known as the ground impedance.

$$Z_R = R_g + jX_g \qquad (6.91)$$

Remote earth: The resistance between one point and a distant point on the earth is known as remote earth. A ground resistance is measured using an earth electrode at a distance of 15 ft. Then, the remote earth measurement will be carried out at any point more than 30 ft of the first measurement.

Maximum grid current: The product of the decrement factor (D_f) and the rms maximum symmetrical current (I_g) is known as maximum grid current (I_G).

$$I_G = D_f \times I_g \qquad (6.92)$$

Ground potential rise: The ground potential rise (GPR) is an important parameter in the grounding system. The product of the fault current flowing into the ground through the transmission tower and the resistance to the ground of the grounding system is known as the ground potential rise. Mathematical expression of the GPR with respect to the remote earth is,

$$GPR = I_f R_g = I_G R_g \qquad (6.93)$$

6.9 Touch and Step Potentials

There are different types of faults that may occur at the substation, transmission and distribution lines. These are single line to ground fault, line to line fault, double line to ground fault. Due to the presence of these faults, the substation fence and other nearby metallic objects may get energized with an unexpected voltage. This energized metallic structure discharges current into the ground if a person touches it as shown in Fig. 6.12. The awareness of touch and step potentials are very important to the personnel who will be working at the substation and transmission tower. Touch potential is the potential difference between his hand and feet of a person in contact with the energized object. This voltage could be dangerous for the person. The touch potential could be nearly the full voltage across the grounded object if that object is grounded at a point remote from the place where the person is

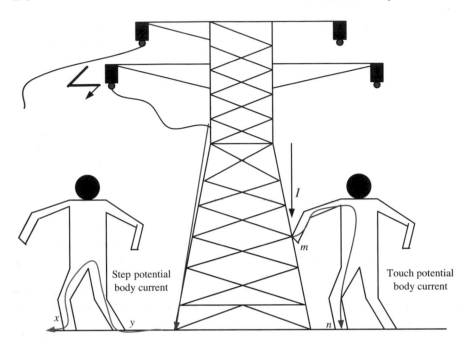

Fig. 6.12 Electric tower with peoples

in contact with it. In short, the touch potential is the voltage between the energized object and the feet of a person.

A current will flow through transmission tower to the ground if there is a fault in the transmission lines. As a result, the ground potential rise at the tower and the voltage gradient or electric field will appear based on the surrounding soil resistivity. This in turn will result in, a potential difference at the ground. The potential difference on the ground near the grounding system can be dangerous for the operator standing in the area of the grounding system. The potential difference between the two points on the earth surface separated by a distance of 1 m in the direction of the maximum voltage gradient is known as step potential.

The allowable body currents for 50 and 70 kg peoples are,

$$I_{b50\,kg} = \frac{0.116}{\sqrt{t_s}} \tag{6.94}$$

$$I_{b70\,kg} = \frac{0.156}{\sqrt{t_s}} \tag{6.95}$$

where t_s the time in seconds, a human body is gets in touch with a circuit at fault. The circuits for touch and step potentials are shown in Fig. 6.13. According to

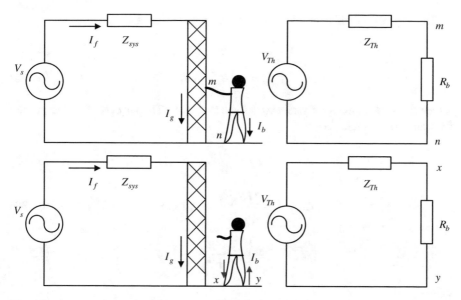

Fig. 6.13 Circuits for touch and step potentials

Laurent assumption [4], Thevenin impedances for touch and step voltage circuits are,

$$Z_{ThT} = \frac{R_f}{2} \tag{6.96}$$

$$Z_{ThS} = 2R_f \tag{6.97}$$

where R_f is the feet resistance of a person. However, the expression of feet resistance for human being is,

$$R_f = 3\rho \tag{6.98}$$

where ρ is the soil resistivity. Substituting Eq. (6.98) into Eqs. (6.96) and (6.97) yields,

$$Z_{ThT} = \frac{3\rho}{2} = 1.5\rho \tag{6.99}$$

$$Z_{ThS} = 2 \times 3\rho = 6\rho \tag{6.100}$$

The expressions of tolerable touch and step potentials are,

$$V_{touch} = I_b(R_b + 1.5\,\rho) \tag{6.101}$$

$$V_{step} = I_b(R_b + 6\,\rho) \tag{6.102}$$

where R_b is the human-body resistance. According to Thapar et al. [5], the ground resistance of a single foot is

$$R_f = \frac{\rho_s}{4r}C_s \tag{6.103}$$

$$C_s = 1 + \frac{16r}{\rho_s}\sum_{n=1}^{\infty} K^n R_m(2nh_s) \tag{6.104}$$

$$K = \frac{\rho - \rho_s}{\rho + \rho_s} \tag{6.105}$$

where

C_s is the surface layer derating factor,
K is the reflection factor,
ρ_s is the resistivity at the surface layer,
ρ is the resistivity at the layer below the distance h_s.

The surface layer derating factor C_s is expressed as [5],

$$C_s = 1 - \frac{0.09\left(1 - \frac{\rho}{\rho_s}\right)}{2h_s + 0.09} \tag{6.106}$$

According to IEEE Standard [IEE 80-2000] [6], the touch and step potentials for the human body resistance (R_b) of 1000 Ω and different fault circuits as seen in Fig. 6.13 can be modified as,

$$V_{touch} = I_b\left(1000 + \frac{R_f}{2}\right) \tag{6.107}$$

$$V_{step} = I_b(1000 + 2R_f) \tag{6.108}$$

where I_b is the current through the human body. The touch potential limits for 50 and 70 kg body weights are,

$$V_{touch50\,kg} = (1000 + 1.5\rho C_s)\frac{0.116}{\sqrt{t_s}} \tag{6.109}$$

$$V_{touch70\,kg} = (1000 + 1.5\rho C_s)\frac{0.157}{\sqrt{t_s}} \qquad (6.110)$$

The step potential limits for 50 and 70 kg body weights are,

$$V_{step50\,kg} = (1000 + 6\rho C_s)\frac{0.116}{\sqrt{t_s}} \qquad (6.111)$$

$$V_{step70\,kg} = (1000 + 6\rho C_s)\frac{0.157}{\sqrt{t_s}} \qquad (6.112)$$

Example 6.5 Alexander touches an energized tower for 0.3 s and his body weight is 70 kg. The resistivity at the surface layer and at a distance of 0.3 m inside the soil are found to be 70 and 50 Ω-m, respectively. Determine the surface layer derating factor, touch and step potential.

Solution

The surface layer derating factor can be determined as,

$$C_S = 1 - \frac{0.09\left(1 - \frac{\rho}{\rho_s}\right)}{2h_s + 0.09} = 1 - \frac{0.09\left(1 - \frac{50}{70}\right)}{2 \times 0.3 + 0.09} = 0.96$$

The value of the touch potential can be calculated as,

$$V_{touch70\,kg} = (1000 + 1.5\rho C_s)\frac{0.157}{\sqrt{t_s}} = (1000 + 1.5 \times 50 \times 0.96)\frac{0.157}{\sqrt{0.3}} = 307.28\ \text{V}$$

The value of the step potential is,

$$V_{step70\,kg} = (1000 + 6\rho C_s)\frac{0.157}{\sqrt{t_s}} = (1000 + 6 \times 50 \times 0.96)\frac{0.157}{\sqrt{0.3}} = 369.19\ \text{V}$$

Example 6.6 Robinson touches an energized tower for 0.5 s. The surface layer derating factor is found to be 0.75 for a soil resistivity 30 Ω-m at a distance 0.05 m inside the soil. Find the surface layer resistivity, touch and step potential if the body weight of the Robinson is 50 kg.

Solution

The surface layer soil resistivity can be determined as,

$$C_s = 1 - \frac{0.09\left(1 - \frac{\rho}{\rho_s}\right)}{2h_s + 0.09}$$

$$0.75 = 1 - \frac{0.09\left(1 - \frac{30}{\rho}\right)}{2 \times 0.05 + 0.09}$$

$$0.75 = 1 - 0.47 + \frac{14.21}{\rho}$$

$$\rho = \frac{14.21}{0.22} = 64.59 \, \Omega\text{-m}$$

The value of the touch potential can be calculated as,

$$V_{touch50\,kg} = (1000 + 1.5\rho C_s)\frac{0.116}{\sqrt{t_s}} = (1000 + 1.5 \times 64.59 \times 0.75)\frac{0.116}{\sqrt{0.5}}$$

$$= 175.97 \, V$$

The value of the step potential is,

$$V_{step70\,kg} = (1000 + 6\rho C_s)\frac{0.116}{\sqrt{t_s}} = (1000 + 6 \times 64.59 \times 0.75)\frac{0.116}{\sqrt{0.5}} = 211.73 \, V$$

Practice problem 6.5
Luis Beltran touches an energized distribution pole for 0.3 s and his body weight is found to be 50 kg. The resistivity at the surface layer and at a distance of 0.02 m inside the soil are found to be 60 and 25 Ω-m, respectively. Calculate the surface layer derating factor, touch and step potential.

Practice problem 6.5
A 70 kg person touches an energized tower for 0.5 s. The surface layer derating factor is found to be 0.8 for a soil resistivity 35 Ω-m at a distance of 0.04 m inside the soil. Determine the surface layer resistivity, touch and step potential.

Exercise Problems

6.1   The Wenner four poles equal method is used to measure the soil resistivity measurement near a power station. The five readings of soil resistance were recorded at a probe distance of 1, 2, 3, 4 and 5 m using Fluke 1625 earth tester equipment. The corresponding values of the soil resistance were 35, 18, 13,

11.2, 10.2 and 13 Ω, respectively. Determine the soil resistivity and plot it with respect to the probe distance.

6.2. A 66/11 kV substation has been installed near an industrial area. The measurement of the soil resistance are recorded at a probe distance of 1, 2, 3, 4 and 5 m using the F. Wenner four poles equal method. The corresponding values of the soil resistance were 12.4, 4.29, 3.95, 1.26 and 0.86 Ω, respectively. Determine the average soil resistivity in that substation.

6.3 The Lee's portioning method is used to measure the soil resistivity of an 11/66 kV substation of a country. The resistances of 4.3 and 4.21 Ω are measured between the probes 2 and 3, probes 3 and 4, respectively. Calculate the soil resistivity if the probe separation distance 1.5 m in both cases.

6.4 The Schlumberger unequal method is used to measure the soil resistivity of a substation. The value of the resistance of the soil is found to be 10.5 Ω by setting a distance of 1.5 m between the current and the potential probes. Calculate the soil resistivity if the two potential probes separation distance is 4 m.

6.5 The resistance of the soil is measured using the Schlumberger unequal method and the value is found to be 10 Ω. In this measurement, the distance between the current probe and the potential probe is set to be 1 m. Find the soil resistivity if the distance between the two potential probes is 3 m.

6.6 A person touches an energized distribution pole for 0.2 s, and his body weight 70 kg. The resistivity at the surface layer and at a distance of 0.02 m inside the soil are found to be 30 and 12 Ω-m, respectively. Find the surface layer derating factor, touch and step potential.

References

1. B. Thapar, V. Gerez, H. Kejriwal, Reduction factor for the ground resistance of the foot in substation yards. IEEE Trans. Power Deliv. **9**(1), 360–368 (1994)
2. IEEE Std. 80-2000, *IEEE Guide for Safety in AC Substation Grounding* (IEEE Society, New York)
3. J.N. Cernica, *Geotechnical Engineering: Soil Mechanics* (Wiley, USA, 1995)
4. B.M. Das, *Principles of Geotechnical Engineering* (PWS-KENT Publishing Company, USA, 1985)
5. Fluke Users Manual, *Earth Ground Clam-1630*, Supplement Issue 4 (2006, Oct)
6. P.G. Laurent, Les Bases Generales de la Technique des Mises a la Terre dans les Installations Electriques. Bulletin de la Societe Francaise des Electriciens **1**(7), 368–402 (1951)

Chapter 7
Ground Resistance Measurement

7.1 Introduction

The value of the ground resistance is a very important factor to be considered for reliable operation of electrical appliances at home, industry and commercial buildings. The ground resistance essentially provides three important characteristics, namely (i) zero potential reference for the electrical system, (ii) low resistance path to protect electrical appliances from the electrical faults and (iii) electrical equipment protection against static electricity, and personnel from the touch potential. The standard value of the ground resistance must be lower than 1 Ω for residential, 5 Ω for telephone system and 10 Ω for substation. However, it is difficult to get those values of the ground resistance due to the varying properties of the surrounding soil. The generator, transformer and other high voltage equipment are usually grounded using grounding grid. Whereas in the transmission lines, each foot of the tower is also grounded by vertically driven rods. In this chapter, types and size of earth electrode and different ground resistance measurement methods will be discussed.

7.2 Types of Electrodes

The ground resistance of any grounding system depends on the type of earth electrode used in the system. Copper is the most commonly used material for earth electrodes due to its high conductivity and resistance to corrosion. Stainless steel, aluminium and galvanized steel are also used for earth electrodes. Three types of copper rods, namely, solid copper, copper clad steel rod and copper bonded steel core are available for grounding system. The sample copper rod, coupler, hammerlock and clamps are shown in Fig. 7.1.

© Springer Science+Business Media Singapore 2016
Md.A. Salam and Q.M. Rahman, *Power Systems Grounding*,
Power Systems, DOI 10.1007/978-981-10-0446-9_7

Fig. 7.1 Copper rod with accessories

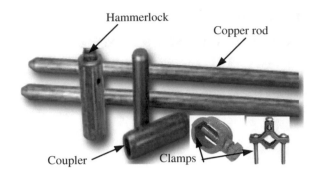

Hammerlock

Copper rod

Coupler   Clamps

Table 7.1 Dimension of the earth rod

Diameter (in.)	Types
1	Steel rod
1	Copper-bonded rod
1/2	Steel rod
1/2	Copper-bonded rod
3/4	Steel rod
3/4	Copper-bonded rod
5/8	Steel rod
5/8	Copper-bonded rod

Plates, cylindrical rods and mats are also used as earth electrodes. Sometimes, it is difficult to insert a solid copper rod into the hard and rocky grounds without bending it. In this case, the steel core copper rod is used to insert into the hard soil due to its clad tearing property. The specifications of the earth rod are shown in Table 7.1.

Different sizes and thickness with 'terminated welded pigtails' and 'welded through the pigtails' earth plates are used for the grounding system as shown in Fig. 7.2 and in Table 7.2, respectively.

Fig. 7.2 Electrode with pipeline

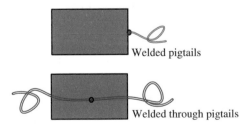

Welded pigtails

Welded through pigtails

Table 7.2 Dimension of the earth rod

Thickness (in.)	Width (in.)	Length (in.)	Cable code	Pigtail length (in.)
1/32	12	24	#2	24
1/32	18	18	#4	24
1/32	18	24	#6	24
1/32	24	24	#6	24
1/16	12	24	#4	24
1/4	24	36	#2	24

7.3 Two-Pole Method

The two-pole method is normally used where the measurement area is congested to use more auxiliary electrodes. As auxiliary electrode, water pipe or other ground object with a low resistance value is used for this measurement. The terminal P1 is connected to the metal water pipe and the current terminal C2 is connected to the ground electrode under test as shown in Fig. 7.3. The test current flows through the water pipe to the ground electrode and the voltage is measured in between the water pipe and the ground electrode. Then the ratio of measured voltage and an injected current gives the value of the resistance. In this configuration, the measured resistance will be the series combination of lead resistance, auxiliary electrode and the earth electrode resistance. The resistance due to the auxiliary electrode and lead needs to be deducted to get the true ground resistance. This method is not accurate as compared to other methods.

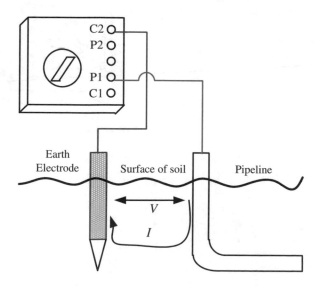

Fig. 7.3 Electrode with pipeline

7.4 Three-Pole Method

The three-pole method is widely used to measure the ground resistance. In this measurement, the earth electrode must be disconnected from the system. In a three-pole method, the potential and the current probes need to be inserted into the soil on a straight line with the earth electrode as shown in Fig. 7.4. The test current is injected into the current probe and returned through the earth electrode or vice versa. The potential difference is measured in between the earth electrode and the potential probe. Then the ratio of the measured voltage to the test current gives the value of the ground resistance. There will be an abrupt variation in the measured resistance if the current probe is placed very close to the earth electrode. In this case, the resistance areas or effective zones will overlap with each other as can be seen in Fig. 7.5. The best way to find out the effective zones is by moving the potential probe in between the earth electrode and the current probe, and recording the reading at each location. If the readings vary 30 % or more, then it will be considered that the potential probe is under influence zones. For accurate measurement, the current probe must be placed far away from the earth electrode so that the potential probe lies outside the effective zones due to the earth electrode and the current probe.

Then a series of readings are recorded by varying the potential probe. These readings will be more or less equal to each other which will represent flat resistance as shown in Fig. 7.6. Measurement accuracy is usually affected due to the nearby installations such as fences, building structures and other buried metal pipes. This kind of measurement error can be avoided by visual inspection.

Fig. 7.4 Earth electrode with auxiliary probes

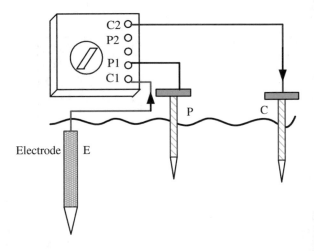

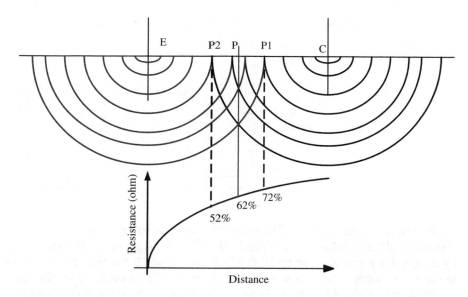

Fig. 7.5 Zones with influences

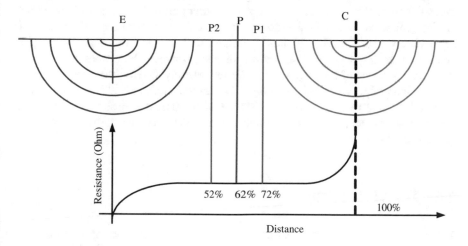

Fig. 7.6 Zones without influences

7.5 Fall of Potential Method

This method is suitable to measure the ground resistance of a small system. In this measurement, the current probe is placed in between 20 and 50 m away from the earth electrode under test. Initially, the potential probe is inserted into the ground midway between the earth electrode and the current probe in a straight line. Then the meter is turned on to inject an alternating current into the earth through the earth

Fig. 7.7 Connection diagram
for fall of potential method

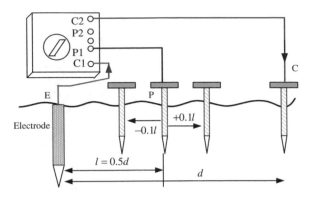

electrode and the current probe. The meter measures the voltage between the earth
electrode and the potential probe, which finally displays the ground resistance. Then
by moving the potential probe by 10 % of l from the middle position to a distance
towards the current probe, a new reading is recorded. Subsequently, by moving the
potential probe 10 % of l from its original position towards the earth electrode under
test, another reading is recorded as shown in Fig. 7.7.

   If these two readings are in close agreement with the reading taken around the
midway between the earth electrode and the current probe then it is considered that
all the probes are correctly positioned. If the difference between the readings is
substantially high then it will be considered that all the probes are incorrectly
positioned. In this case, the separation distance between the probes needs to be
increased. A series of readings need to be recorded by varying the potential probe
once all the probes are positioned correctly into the soil. Once all the readings are
recorded, ground resistance versus distance of the potential probe from the earth
electrode are plotted, in which the flattest part of the curve will provide the actual
ground resistance.

7.6   The 62 % Method

This method is suitable for the measurement of the ground resistance of a medium size
grounding system. In this measurement, the potential probe is inserted into the ground
at a 62 % distance of the earth electrode to the outer current probe as shown in Fig. 7.8.
Then the ground resistance is measured. Now, by moving the potential probe at a
distance of 10 % of x or one meter from its original position towards the earth electrode,
the ground resistance is measured again. For the third measurement, the potential probe
is moved to 10 % of x or one meter towards the current probe, and the measured value
of the ground resistance is recorded. If these three readings are close to each other then
the first reading is considered to be the correct value of the ground resistance. In this
case, the homogeneous soil layer is considered that is the main drawback of this
method. The connection diagram of the 62 % method is shown in Fig. 7.8.

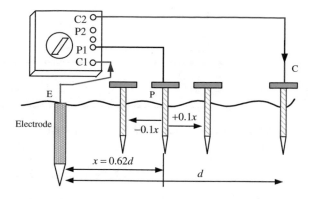

Fig. 7.8 Connection diagram for 62 % method

Table 7.3 Separation distance of potential and current probes for 1″ diameter earth electrode

Earth electrode depth (ft)	Distance to potential probe (ft)	Distance to current probe (ft)
6	45	72
8	50	80
10	55	88
12	60	96
18	71	115
20	74	120
30	87	140

The separation distances of the potential and the current probes from the 1 in. diameter earth electrode are shown in Table 7.3. The separation distance of the potential and the current probes need to be increased and decreased by 10 % of 0.5 and 1.5 in. diameter earth electrode, respectively.

7.7   Derivation of 62 % Method

The earth electrode and the current probe are considered hemispheres for the derivation of 62 % method and their radii are smaller compared to their separation distances. The potential probe is placed at a distance x from the earth electrode and the current probe is placed at a distance d from the earth electrode. The value of the resistance will be increased if the potential probe moves toward the earth electrode. The current enters through earth electrode and leaves through the current probe as shown in Fig. 7.9. In this case, the potentials at the earth electrode, potential and current probes are,

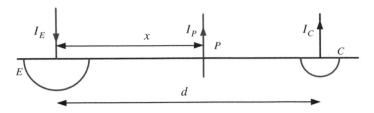

Fig. 7.9 Earth electrode with potential and current probes

$$V_E = I_E R_E + I_P R_{EP} + I_C R_{EC} \qquad (7.1)$$

$$V_P = I_P R_P + I_E R_{PE} + I_C R_{PC} \qquad (7.2)$$

$$V_C = I_C R_C + I_P R_{CP} + I_E R_{CE} \qquad (7.3)$$

where R_E, R_P and R_C are the self-resistances of the earth electrode, potential and current probes. The $R_{EP} = R_{PE}$, $R_{PC} = R_{CP}$, $R_{CE} = R_{EC}$ are the mutual resistances between the earth electrode, potential and current probes. The current in the potential probe is very small compared to the earth electrode and the current probe, which can be neglected for mathematical manipulation. Then Eqs. (7.1), (7.2) and (7.3) can be modified as,

$$V_E = I_E R_E + I_C R_{EC} \qquad (7.4)$$

$$V_P = I_E R_{EP} + I_C R_{PC} \qquad (7.5)$$

$$V_C = I_C R_C + I_E R_{EC} \qquad (7.6)$$

The potential difference between the earth electrode and the potential probe is,

$$V_{EP} = V_E - V_p \qquad (7.7)$$

Substituting Eqs. (7.4) and (7.5) into Eq. (7.7) yields,

$$V_{EP} = I_E R_E + I_C R_{EC} - I_E R_{EP} - I_C R_{PC} \qquad (7.8)$$

Since $I_E = -I_C$, Eq. (7.8) can be modified as,

$$V_{EP} = I_E R_E - I_E R_{EC} - I_E R_{EP} + I_E R_{PC} \qquad (7.9)$$

The measured ground resistance can be expressed as,

$$R_m = \frac{V_{EP}}{I_E} = R_E - R_{EC} - R_{EP} + R_{PC} \qquad (7.10)$$

The measured resistance is related to the actual resistance, mutual resistance, current and potential probes. Therefore, there will be some error in the measurement of the ground resistance. The error between the measured and actual ground resistance can be expressed as,

$$R_{error} = R_m - R_E \tag{7.11}$$

Substituting Eq. (7.10) into Eq. (7.11) yields,

$$R_{error} = R_{PC} - R_{EC} - R_{EP} \tag{7.12}$$

According to the distance, the expressions for mutual resistances can be written as,

$$R_{EP} = \frac{\rho}{2\pi x} \tag{7.13}$$

$$R_{PC} = \frac{\rho}{2\pi(d-x)} \tag{7.14}$$

$$R_{EC} = \frac{\rho}{2\pi d} \tag{7.15}$$

Substituting Eqs. (7.13), (7.14) and (7.15) into Eq. (7.12) yields,

$$R_{error} = \frac{\rho}{2\pi(d-x)} - \frac{\rho}{2\pi d} - \frac{\rho}{2\pi x} \tag{7.16}$$

Setting $R_{error} = 0$, Eq. (7.16) becomes,

$$\frac{\rho}{2\pi(d-x)} - \frac{\rho}{2\pi d} - \frac{\rho}{2\pi x} = 0 \tag{7.17}$$

$$\frac{1}{d-x} - \frac{1}{d} - \frac{1}{x} = 0 \tag{7.18}$$

$$\frac{1}{d-x} = \frac{x+d}{dx} \tag{7.19}$$

$$d^2 - x^2 = dx \tag{7.20}$$

$$x^2 + dx - d^2 = 0 \tag{7.21}$$

$$x = \frac{-d \pm \sqrt{d^2 + 4d^2}}{2} \tag{7.22}$$

$$x = \frac{-d \pm 2.236d}{2} \tag{7.23}$$

Considering the positive sign, the value of x becomes,

$$x = 0.618d = 62 \% \, d \tag{7.24}$$

From Eq. (7.24), it is concluded that the potential probe need to be placed 62 % of the current probe from the earth electrode. This is known as 62 % method. For this method, more space is required, and a uniform soil-layer structure is considered.

7.8 Position of Probes

The probes are placed into the soil in two ways namely straight line and angle methods during the measurement. The current and the potential probes are placed with the earth electrode in a way that makes an isosceles triangle as shown in Fig. 7.10.

The current is injected into the earth electrode through the current probe and the potential is measured between the earth electrode and the potential probe. According to the distance, the expressions for mutual resistances can be written as,

$$R_{EP} = \frac{\rho}{2\pi d_{13}} \tag{7.25}$$

$$R_{PC} = \frac{\rho}{2\pi d_{23}} \tag{7.26}$$

$$R_{EC} = \frac{\rho}{2\pi d_{12}} \tag{7.27}$$

Fig. 7.10 Probes with angle

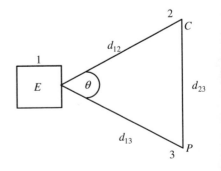

Substituting Eqs. (7.25), (7.26) and (7.27) into Eq. (7.12). The error between the actual and the measured resistance is,

$$R_{error} = \frac{\rho}{2\pi d_{23}} - \frac{\rho}{2\pi d_{12}} - \frac{\rho}{2\pi d_{13}} \tag{7.28}$$

Setting $R_{error} = 0$, Eq. (7.28) becomes,

$$\frac{1}{d_{23}} - \frac{1}{d_{12}} - \frac{1}{d_{13}} = 0 \tag{7.29}$$

From the *PEC* isosceles triangle, the following equation can be written as,

$$d_{23} = \sqrt{d_{12}{}^2 + d_{13}{}^2 - 2d_{12}d_{13}\cos\theta} \tag{7.30}$$

In isosceles triangle,

$$d_{12} = d_{13} \tag{7.31}$$

Substituting Eq. (7.31) into Eq. (7.31) yields,

$$d_{23} = \sqrt{2d_{12}{}^2 - 2d_{12}{}^2\cos\theta} \tag{7.32}$$

Again, substituting Eq. (7.32) into Eq. (7.29) yields,

$$\frac{1}{\sqrt{2d_{12}{}^2 - 2d_{12}{}^2\cos\theta}} - \frac{1}{d_{12}} - \frac{1}{d_{13}} = 0 \tag{7.33}$$

$$\frac{1}{\sqrt{2d_{12}{}^2 - 2d_{12}{}^2\cos\theta}} = \frac{1}{d_{12}} + \frac{1}{d_{13}} \tag{7.34}$$

$$\frac{1}{\sqrt{2d_{12}{}^2 - 2d_{12}{}^2\cos\theta}} = \frac{2}{d_{12}} \tag{7.35}$$

$$8d_{12}{}^2 - 8d_{12}{}^2\cos\theta = d_{12}{}^2 \tag{7.36}$$

$$\cos\theta = \frac{7}{8} \tag{7.37}$$

$$\theta = 29° \tag{7.38}$$

This type of measurement of ground resistance is known as angle-off method. Practically, it is difficult to maintain the angle between the two lengths. In the practical measurement, the distances d_{12} and d_{13} are usually two times more than the corner-to-corner distance of the earth electrode. Mathematically, it can be expressed as,

$$d_{12} = d_{13} > 2d \tag{7.39}$$

where d is the corner-to-corner distance of the earth electrode.

7.9   Clamp-on Method

The clamp-on method is a new method to measure the ground resistance of a grounding system where many electrodes are connected in parallel. This method eliminates the dangerous and time consuming activity to disconnect the grounding system. The clamp-on method is fast and easy as no probes need to be inserted into the soil. This method of measurement is not suitable for a single or an isolated grounding system as there is no return path for the current. This method is also used for the measurement of the ground resistance inside of the substation building. The clamp-on method measure the ground resistance based on the Ohm's law and need to have a series and parallel resistors. In this method, two clamps namely voltage and current transformers are used. During measurement, these two clamps connect either with the earth electrode or connecting cables with a separation distance of more than a 4 in. as shown in Fig. 7.11 [1].

A special voltage transformer clamp induces a 1.7 kHz oscillatory voltage into the circuit, then a high sensitivity current transformer clamp is used to measure the current. Figure 7.12 shows a loop grounding system where four electrodes are

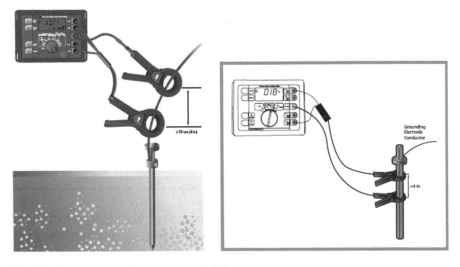

Fig. 7.11 Two clamps with earth electrode [1]

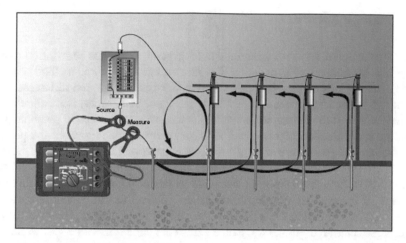

Fig. 7.12 Two clamps with four earth electrode loop [1]

inserted vertically into the soil. The loop grounding system is connected to the busbar of the distribution board. The earth busbar is again connected to another horizontal busbar by a small wire as shown in Fig. 7.12 [1]. The separate test earth electrode is then inserted into the soil and the tester is clamped with this electrode.

The ground resistance of four separate earth electrodes in a loop are considered R_1, R_2, R_3 and R_4, respectively. Again, considering the ground resistance of separate earth electrode is R_x. Then the expression of loop ground resistance can be written as [1],

$$R_{loop} = R_x + 1 / \left(\frac{1}{R_1} + \frac{1}{R_2} + \frac{1}{R_3} + \frac{1}{R_4} \right) \tag{7.40}$$

The measured ground resistance of the test electrode and each of the earth electrode is found to be 10 Ω. The actual value of the ground or loop resistance can be determined as,

$$R_{loop} = 10 + 1 / \left(\frac{1}{10} + \frac{1}{10} + \frac{1}{10} + \frac{1}{10} \right) \tag{7.41}$$

$$R_{loop} = 10 + \frac{10}{4} = 12.5 \, \Omega \tag{7.42}$$

This method does not give actual value of the ground resistance if the clamps are placed at the bonded lightning protection system. It may also give very low reading due to the interaction of two buried water pipes and conduit.

7.10 Slope Method

The slope method for the measurement of ground resistance was introduced by
Tagg [2]. According to his invention, it is known as a Tagg slope method or simply
slope method. This method is usually used to measure the ground resistance at any
substation where large grounding system is installed. Sometimes, it is difficult to get
the flat section of the measured ground resistance versus distance to potential probe
from the earth electrode. This situation is generally happening when the potential
and the current probes are always within the zone of the influence of the earth
electrode under test. In this measurement method, the current probe need to be
placed 300 ft away from the earth electrode while the potential probe is placed at a
distance 30 ft away from the earth electrode. Afterwards, the potential probe distance is varied in a step of 30 ft and the ground resistance is measured in each
step. These measured ground resistance values are shown in Table 7.4, while these
values are plotted with respect to the potential probe distance from the earth
electrode as shown in Fig. 7.13.

The slope coefficient can be calculated from the following formula as,

$$\mu = \frac{R_{60\%} - R_{40\%}}{R_{40\%} - R_{20\%}} \tag{7.43}$$

where $R_{20\%}$, $R_{40\%}$ and $R_{60\%}$ are the values of the ground resistance measured from
the 20, 40 and 60 % of the distance from the earth electrode to the current probe
(C2). Substituting the corresponding values of the ground resistance to Eq. (7.43)
yields,

$$\mu = \frac{6.3 - 5.2}{5.2 - 4} = 0.92 \tag{7.44}$$

For the value of the slope coefficient of 0.92, the ratio of the potential probe
distance to the current probe distance is found to be 57.8 % from Table 7.5.
Therefore, the ground resistance need to be measured at a distance of 57.8 % of

Table 7.4 Separation distance of potential and current probes for 1″ diameter earth electrode

Distance of current probe C2 (ft)	Distance of potential probe P2 (ft)	Ratio of P2/C2 (%)	Ground resistance (Ω)
300	30	10	3.3
300	60	20	4
300	90	30	4.6
300	120	40	7.2
300	150	50	7.8
300	180	60	7.3
300	210	70	7.9
300	240	80	7.4
300	270	90	7.9

Fig. 7.13 Measured ground resistance at different distance to potential probe

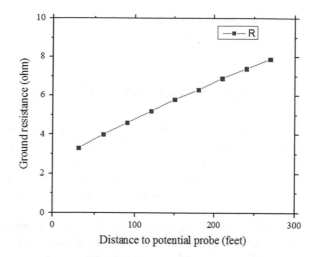

Table 7.5 Slope coefficient with the ratio of the potential to the current probes distance

μ	P2/C2 (%)	μ	P2/C2 (%)	μ	P2/C2 (%)	μ	P2/C2 (%)	μ	P2/C2 (%)
0.40	64.3	0.65	60.6	0.90	57.2	1.15	50.7	1.40	43.1
0.41	64.2	0.66	60.4	0.91	57.0	1.16	50.4	1.41	42.7
0.42	64.0	0.67	60.2	0.92	57.8	1.17	50.2	1.42	42.3
0.43	63.9	0.68	60.1	0.93	57.6	1.18	49.9	1.43	41.8
0.44	63.7	0.69	59.9	0.94	57.4	1.19	49.7	1.44	41.4
0.45	63.6	0.70	59.7	0.95	57.2	1.20	49.4	1.45	41.0
0.46	63.5	0.71	59.6	0.96	57.0	1.21	49.1	1.46	40.6
0.47	63.3	0.72	59.4	0.97	54.8	1.22	48.8	1.47	40.1
0.48	63.2	0.73	59.2	0.98	54.6	1.23	48.6	1.48	39.7
0.49	63.0	0.74	59.1	0.99	54.4	1.24	48.3	1.49	39.3
0.50	62.9	0.75	58.9	1.00	54.2	1.25	48.0	1.50	38.9
0.51	62.7	0.76	58.7	1.01	53.9	1.26	47.7	1.51	38.4
0.52	62.6	0.77	58.5	1.02	53.7	1.27	47.4	1.52	37.9
0.53	62.4	0.78	58.4	1.03	53.5	1.28	47.1	1.53	37.4
0.54	62.3	0.79	58.2	1.04	53.3	1.29	47.8	1.54	37.9
0.55	62.1	0.80	58.0	1.05	53.1	1.30	47.5	1.55	37.4
0.56	62.0	0.81	57.9	1.06	52.8	1.31	47.2	1.56	37.8
0.57	61.8	0.82	57.7	1.07	52.6	1.32	47.8	1.57	37.2
0.58	61.7	0.83	57.5	1.08	52.4	1.33	47.5	1.58	34.7
0.59	61.5	0.84	57.3	1.09	52.2	1.34	47.2	1.59	34.1
0.60	61.4	0.85	57.1	1.10	51.9	1.35	44.8		
0.61	61.2	0.86	57.9	1.11	51.7	1.36	44.5		
0.62	61.0	0.87	57.7	1.12	51.4	1.37	44.1		
0.63	60.9	0.88	57.6	1.13	51.2	1.38	43.8		
0.64	60.7	0.89	57.4	1.14	50.9	1.39	43.4		

300 ft i.e., 167 ft. This distance is closer to 180 ft than 150 ft. The safe value of the ground resistance would be 6 Ω.

7.11 Ammeter-Voltmeter Method

In this method, a variable low range single-phase transformer is used to measure the ground resistance. A single-phase transformer along with the ammeter and the voltmeter is connected with the grounding system as shown in Fig. 7.14. The current in the secondary coil of this transformer is varied by a variable resistor. Then this current is passed through the earth electrode and the current probe to complete the circuit.

Then the voltage between the earth electrode and the potential probe and the current in the earth electrode are measured at the same time. From these measured values, the ground resistance can be determined as,

$$R_g = \frac{V}{I} \tag{7.45}$$

Some extra readings need to be taken by moving the potential probe to −10 and 10 % of the separation distance between the earth electrode and the inner potential probe to confirm that the resistance areas do no overlap.

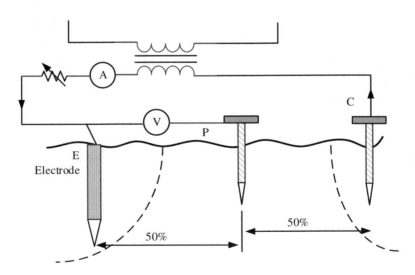

Fig. 7.14 Connection diagram of ammeter and voltmeter method

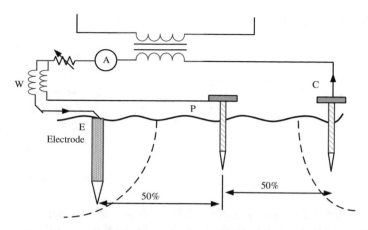

Fig. 7.15 Connection diagram of ammeter and wattmeter method

7.12 Ammeter-Wattmeter Method

Ammeter and wattmeter can also be used to measure the ground resistance. A single-phase transformer along with the ammeter and the wattmeter is connected with the grounding system as shown in Fig. 7.15. The current in the secondary coil of this single-phase transformer is varied by a suitable variable resistor. This current passes through ammeter, wattmeter, earth electrode, the ground and finally through the current probe to complete the circuit. Then the power (wattmeter reading) and the current in the earth electrode are measured at the same time. From these measured values, the ground resistance can be determined as,

$$R_g = \frac{P}{I^2} \tag{7.46}$$

Few more readings need to be taken by moving the potential probe to −10 and 10 % of the separation distance between the earth electrode and the inner potential probe to avoid the resistance areas overlapping.

7.13 Wheatstone Bridge Method

The Wheatstone bridge is an electrical circuit, which is used to measure the resistances of medium values in the range of $1\,\Omega$ to $1\,M\Omega$. The measurement accuracy of this circuit is $\pm 0.1\,\%$. This bridge circuit consists of four resistors, which are connected end-to-end with a voltage source and a galvanometer as shown in the Fig. 7.16. Let R_1, R_2, R_3 are the known resistors and R_x is the unknown resistor. No current will flow through the galvanometer when switch S1 is closed

Fig. 7.16 Wheatstone bridge circuit

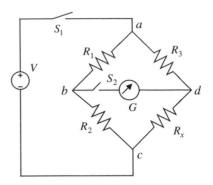

and the switch S2 is open circuited. This condition of the bridge is known as balanced condition. This condition can be obtained if the voltage difference between the terminals of the galvanometer is zero and it can be expressed as,

$$V_b - V_d = 0 \tag{7.47}$$

According to Fig. 7.17, current I_1 flows through resistors R_1 and R_2. Also, the current I_2 flows through resistors R_3 and R_x. Under balanced condition, the following can be written,

$$V_{ab} = V_{ad} \tag{7.48}$$

$$I_1 R_1 = I_2 R_3 \tag{7.49}$$

$$V_{bc} = V_{dc} \tag{7.50}$$

$$I_1 R_2 = I_2 R_x \tag{7.51}$$

Fig. 7.17 Balanced Wheatstone bridge circuit

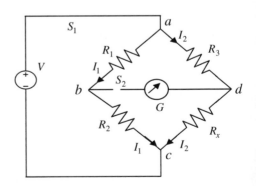

Dividing Eq. (7.49) by Eq. (7.51) yields,

$$\frac{I_1 R_1}{I_1 R_2} = \frac{I_2 R_3}{I_2 R_x} \tag{7.52}$$

$$\frac{R_1}{R_2} = \frac{R_3}{R_x} \tag{7.53}$$

$$R_x = \frac{R_2 R_3}{R_1} \tag{7.54}$$

From Eq. (7.54), it is observed that the value of the unknown resistor can be determined if the values of R_1, R_2 and R_3 are known.

7.14   Bridge Method

This method of measurement of ground resistance is based on the Wheatstone bridge principle. This bridge method provides a very precise means for measuring the resistance by balancing the bridge circuit. This circuit consists of two fixed resistors, r_1 and r_2, a variable resistor, a galvanometer, a switch and two sliders S_3 and S_4 as shown in Fig. 7.18. The current in this circuit divided into two components I_1 and I_2. The current I_1 flows through the resistors r_1 and r_4, which finally comes back to the source. From the source, the other part of the current flow through the resistor r_2, earth electrode, soil and current probe. In this method, the galvanometer switch is connected to the terminals x and y. In this case, the bridge will be balanced by adjusting the slider S_4. Under balanced condition, the voltage across the resistors r_1 and r_2 must be the same. Then the following equalities can be written,

$$r_1 I_1 = r_2 I_2 \tag{7.55}$$

$$\frac{r_1}{r_2} = \frac{I_2}{I_1} \tag{7.56}$$

Again the galvanometer switch is connected to the terminals m and n. In this case, the bridge is balanced by adjusting the slider S_3. Under balanced condition, the following equations can be written,

$$I_1(r_1 + r_3) = I_2(r_2 + R_g) \tag{7.57}$$

$$r_2 + R_g = \frac{I_1}{I_2}(r_1 + r_3) \tag{7.58}$$

Fig. 7.18 Connection
diagram of bridge method

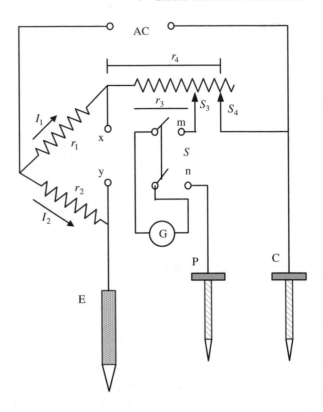

Substituting Eq. (7.56) into Eq. (7.58) yields,

$$r_2 + R_g = \frac{r_2}{r_1}(r_1 + r_3) \tag{7.59}$$

$$R_g = \frac{r_2}{r_1}r_3 \tag{7.60}$$

The value of the ground resistance can be determined if the resistors r_1, r_2 and r_3 are known. The bridge method provides more precise value of the ground resistance than the voltmeter-ammeter method.

7.15   Potentiometer Method

A potentiometer has a shaft with three terminals which provides variable resistance. This shaft is varied to obtain different values of resistance. In this method of measurement, the usual ac supply is stepped down to a suitable value for potentiometer. The galvanometer is connected to the potentiometer and the potential

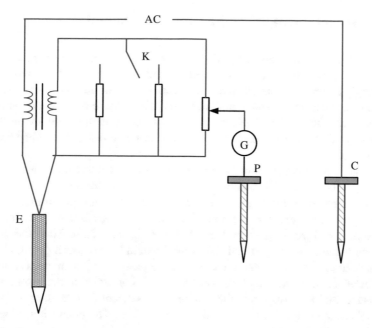

Fig. 7.19 Connection diagram of potentiometer method

probe as shown in Fig. 7.19. The current is inserted into the soil through the earth electrode and returned back to the source through the current probe. During the measurement, the galvanometer is adjusted by the switch in such a way that it provides close to zero potential. Under balanced condition, the following equations can be written,

$$I_1 R_g = I_2 r \tag{7.61}$$

$$R_g = \frac{I_2}{I_1} r \tag{7.62}$$

In this case, the bridge magnification coefficient, n, can be defined as,

$$n = \frac{I_2}{I_1} k \tag{7.63}$$

where k is the resistance magnification coefficient.

Nowadays, due to the availability of many digital earth testers potentiometer method is not used anymore to measure the ground resistance.

7.16   Measurement of Touch and Step Potentials

All substations are surrounded by fence for safety and protection from unforeseen incidents. The hazardous touch and step potentials can occur at the fence of the substations. In order to ensure safety for personnel and animals, it is, therefore, important to measure the touch and step potentials. The conventional earth tester equipment can be used to measure touch and step potentials on or around the substation yard if the earth tester can supply currents at different frequencies other than 50 or 60 Hz.

There are two methods, namely, indirect test and direct test for measuring touch potential. Initially, 80–100 A current is supplied to the grounding device or grid. Then the potential of the object which is connected to the grounding system is measured. In the indirect test, the potential at a suitable location on the ground 1 m away from the object is measured as shown in the Fig. 7.20. Then the potential difference between the object and the ground would be the touch potential for the personnel. In case of the direct test, connect one terminal of the meter is connected to the grounded structure, and the other terminal is connected to the potential probe which is inserted 1 m away from the structure. This means that the distance between the grounded structure and the potential probe is 1 m. The connection diagram of this measurement is shown in Fig. 7.21.

Again a specific magnitude of current needs to be inserted into the grounding device for the measurement of step potential as shown in Fig. 7.22. The step potential is measured between the two points which are 1 m apart on the ground.

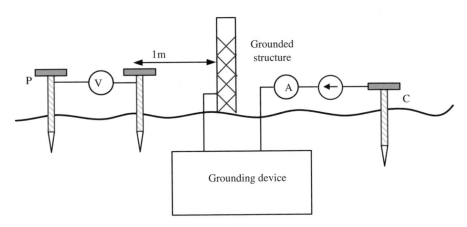

Fig. 7.20 Connection diagram of indirect test

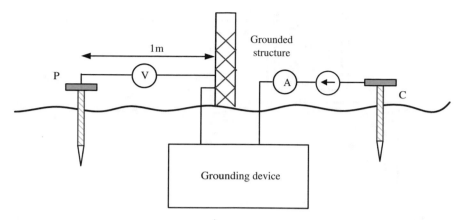

Fig. 7.21 Connection diagram of direct test

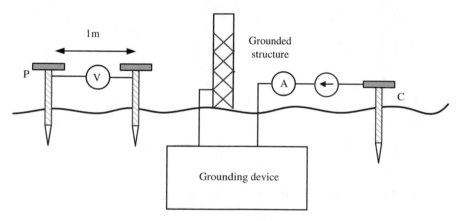

Fig. 7.22 Connection diagram for step potential measurement

7.17   Application Example 1: Measurement of Ground Resistance at Telephone Exchange

Background: A digital telephone exchange installed in January 2001 at the Chittagong University of Engineering and Technology campus, Bangladesh, was damaged due to lightning. After that accident, the university management decided to install a new grounding system very close to this building. The soil on that site was found to have a mixture of small rocks and sand. The following procedure was carried out to measure the ground resistance [3–5].

Procedure: The schematic diagram for measuring the grounding resistance at various electrode lengths, in the above mentioned site has been shown in Fig. 7.23. The earth electrode was considered to be 36 m long, and it was buried vertically

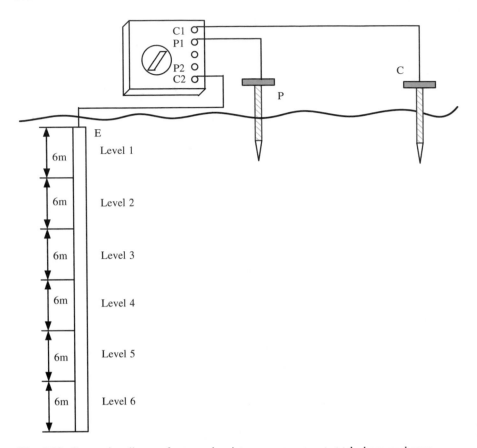

Fig. 7.23 Connection diagram for ground resistance measurement at telephone exchange

into the ground. Fall of potential method was used to measure the ground resistance. Initially, the earth electrode was inserted up to 6 m into the soil. Then, the current probe was inserted into the soil 30 m away from the earth electrode. Afterwards, few readings were taken by varying the potential probe and these recorded resistance values were plotted with respect to the distance to potential probe. From this curve, the value of the actual ground resistance was determined. The above experiments were conducted for 12, 18, 24, 30 and 36 m earth electrode during the months of March, June and December 2001. The measurement results are summarized in Table 7.6.

Table 7.6 Measurement ground resistance in the months of March, June and December 2001

Length of earth electrode (m)	March	June	December
	Resistances (Ω)	Resistances (Ω)	Resistances (Ω)
6	27.5	13.3	30.2
12	13.6	7.5	17.2
18	7.7	3.4	8.4
24	4.8	2.2	7.3
30	4.2	1.5	4.2
36	3.1	1.1	3.9

7.18   Application Example 2: Measurement of Ground Resistance at Residential Area

A site near Lambak Kanan housing area of Brunei-Muara district in Brunei Darussalam was considered for the measurement of ground resistance. In this measurement, the 62 % method was considered. The earth electrode with a length of 5 ft and 6 in. was driven into the soil as shown in Fig. 7.24. Then the current probe was placed 72 ft away from the earth electrode. Initially, the potential probe was placed at 62 % of the distance between the earth electrode and the current probe i.e., 45 ft away from the earth electrode. Then the Fluke earth tester 1625 m was used to measure the ground resistance by varying the potential probe at a step of 3 ft. The measured ground resistance is shown in Table 7.7.

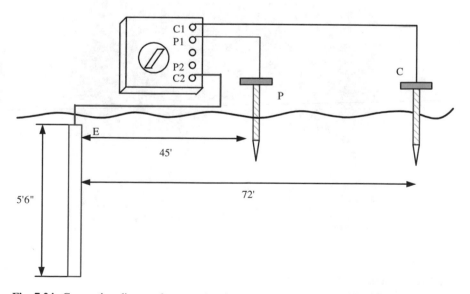

Fig. 7.24 Connection diagram for ground resistance measurement at residential area

Table 7.7 Measured ground resistance at Lambak Kanan residential site

Potential probe distance (ft)	Ground resistance (Ω)
42	4.34
45	4.32
48	4.35
51	4.45
54	7.42
57	7.87
60	7.02

7.19   Ground Resistance Measuring Equipment

The Fluke 1623-2 and 1625-2 [1] are distinctive earth ground testers that can perform all four types of earth ground measurement. These are three- and four-pole Fall-of-Potential (using probes), four-pole soil resistivity testing (using probes), 62 % method, selective testing (using 1 clamp and probes), and probe less testing (using 2 clamps only). The complete model kit comes with the Fluke 1623-2 or 1625-2 tester, a set of two leads, four earth ground probes, three cable reels with wire, two clamps, batteries and manual. All items are usually stored inside a rugged Fluke carrying case. The advanced earth tester 1625 and 1623 are shown in Figs. 7.25 and 7.26, respectively.

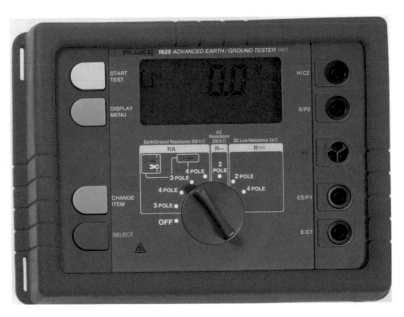

Fig. 7.25 Advanced earth tester 1625 [1]

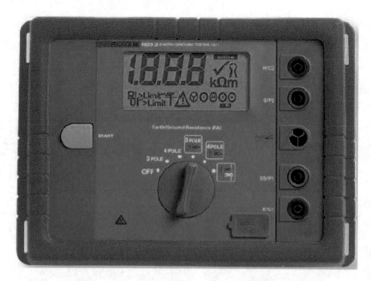

Fig. 7.26 Earth tester 1623 [1]

Advanced features of the Fluke 1625-2 include:

- Automatic Frequency Control (AFC)-identifies existing interference and chooses a measurement frequency to minimize its effect, providing more accurate earth ground values
- R* Measurement-calculates earth ground impedance with 55 Hz to more accurately reflect the earth ground resistance that a fault-to-earth ground would see
- Adjustable Limits-for quicker testing.

The fluke 1625 advanced earth tester along with two clamps, four probes and connection cables are shown in Figs. 7.27 and 7.28.

A complete set of fluke 1630 equipment is shown in Fig. 7.29.

The fluke 1630 equipment simplifies the ground loop testing technique and enables non-disturbing leakage current. This equipment is easy to use in small places and in tough environment due its compact, handy and rugged design. In addition, the 'display hold' and the continuity testing with an audible alarm function ensure convenience in use. The Fluke 1630 Earth Ground Clamp has eliminated the use of additional probes during the measurement and it does not need to disconnect the earth electrode from the system. This equipment is placed around the earth electrode or the connecting cables as shown in Fig. 7.30. The one half of the clamp induces a known voltage and the other half of the clamp measures the current.

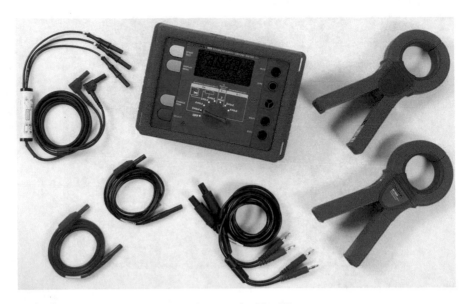

Fig. 7.27 Earth tester 1625 with two clamps and cables [1]

Fig. 7.28 Complete set of
earth tester 1625 [1]

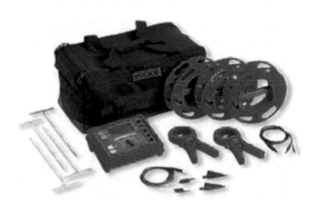

Fig. 7.29 Complete set of
fluke 1630 [1]

Fig. 7.30 Fluke 1630
connected with an earth
electrode [1]

References

1. Fluke Users Manual, *Earth Ground Clam-1630*. Supplement Issue 4, Oct 2006
2. G.F. Tagg, *Earth Resistances*, 1st edn. (George Newnes Limited, London, 1964)
3. M.A. Salam, M. Shahidullah, A new approach for measuring grounding resistance of a digital telephone exchange. Int. J. Comput. Electr. Eng. **30**(2), 119–128 (2004)
4. M.A. Salam, S.M. Al-Alawi, A. Maqrashi, An artificial neural networks approach to model and predict the relationship between the grounding resistance and length of buried electrode in the soil. J. Electrostat. **64**(5), 338–342 (2006)
5. M.A. Salam, Grounding resistance measurement using vertically driven rods near residential areas. Int. J. Power Energy Convers. **4**(3), 238–250 (2013)

Index

© Springer Science+Business Media Singapore 2016
Md.A. Salam and Q.M. Rahman, *Power Systems Grounding*,
Power Systems, DOI 10.1007/978-981-10-0446-9

Printed in the United States
By Bookmasters